AF352209

COMMERCIAL ORGANIC FLOCCULANTS

COMMERCIAL
ORGANIC FLOCCULANTS

Josef Vostrčil
František Juračka

NOYES DATA CORPORATION

Park Ridge, New Jersey, U.S.A.

1976

Published in the United States of America by
Noyes Data Corporation
Noyes Building, Park Ridge, New Jersey 07656

FOREWORD

A wide range of organic polymers is now being used for water and wastewater treatment under the collective names of flocculants or coagulants and polyelectrolytes.

These chemicals promote the process whereby suspended solids and colloidal materials in the water are agglomerated by chemical means into masses sufficiently large to settle. Being polyelectrolytes, they neutralize the charge on wet dirt particles, allowing them to come together.

Organic flocculants are practically nontoxic, biodegradable, small in volume, easily incinerated, and effective under varied pH and temperature conditions. This makes them ideal for clarification of drinking water from reservoirs, etc. Municipalities use more flocculants than industry and more flocculants are consumed for water clarification than wastewater treatment. However, there is now increased usage for secondary treatment of sewage in competition with biological processes. Also many industrial effluents are detoxified best with organic polyelectrolytes.

In order to facilitate the use of these chemicals the world over, this book presents in tabular form the commercial sources, availability, and major applications of organic flocculants on a worldwide basis. This includes data on chemical composition, patent reviews, and a very complete listing of trademarks, generic, and chemical names.

Advanced composition and production methods developed by Noyes Data are employed to bring our new durably bound books to you in a minimum of time. Special techniques are used to close the gap between "manuscript" and "completed book." Industrial technology is progressing so rapidly that time-honored, conventional typesetting, binding and shipping methods are no longer suitable. We have bypassed the delays in the conventional book publishing cycle and provide the user with an effective and convenient means of reviewing up-to-date information in depth.

CONTENTS

INTRODUCTION

Industrial development accompanied by an increase in living standards results not only in an increase in consumption of pure water but also in an increase in its contamination. Since the sources of water of suitable quality are limited, the pressing need for increasing the available amounts of good quality water spurs the development of new technological processes for water treatment and purification. The use of high molecular weight synthetic organic compounds in a water economy represents one such intensification and rationalization process.

In practice, these high molecular weight organic compounds are given group names according to their mode of action and application. By organic flocculants (polymeric flocculants, polycoagulants) is meant that group of organic compounds, mostly of high molecular weight, which cause the destabilization of suspensions for the most part by forming bridges between dispersed solids particles (bridge-forming flocculants). In the water treatment field these compounds are generally called polyelectrolytes. Many flocculants are electrolytes, although nonionic hydrophilic colloids with flocculating properties similar to those of polyelectrolytes are also known. Furthermore, selective organic flocculants may also be used in the separation processes of mineral mixtures, because under certain conditions the organic flocculant is adsorbed onto the surface of only one mineral.

These organic high molecular weight compounds can be somewhat further classified according to their application as follows.

Polycoagulants: That group of organic high molecular weight compounds causing the destabilization of suspensions, for the most part by reduction of the electrokinetic potential of the surface of the particles, by coagulation. In contrast to inorganic coagulants, polycoagulants also simultaneously act as a bridging mechanism for the polymer chains in addition to reducing electrokinetic potential.

Coagulation Aids: Organic polymeric compounds used along with inorganic coagulants.

Weighting Agents (Sedimentation Aids): Organic compounds which act mostly by increasing the specific gravity of a suspension and/or increasing the settling velocity.

Filtration Aids: Organic polymeric compounds which act to improve the efficiency of the filtration process, whether by improvement of filtration properties of suspensions or by the modification of the filter medium.

Organic Dispersing Agents (Dispersants, Deflocculants): In cases where, in addition to their flocculating effect, use is made of the stabilizing-dispersing effect of organic high molecular weight compounds (sometimes accomplished merely by changing their dosage or molecular weight).

Organic Dewatering and/or Organic Conditioning Aids: High molecular weight organic compounds employed for the improvement of dewatering properties of industrial suspensions and waste sludges.

Corrosion Inhibitors: Organic compounds used to eliminate water corrosion.

Demulsifying Agents: High molecular weight compounds used for breaking oil-in-water emulsions.

The criteria for successful utilization of high molecular weight organic, water-soluble compounds vary with their application. These criteria are influenced either by the choice of the type and the molecular weight of the high molecular weight compound or by the choice of the proper dosage of the polymer in its various applications. Thus the same high molecular weight compound may be classified in several different groups.

Water-soluble organic polymers, especially synthetic ones, have brought about great changes in water technology. Not only do they contribute to the economical operation of many water treatment plants, but they also manage to solve, in some cases, specific problems in water economy. Increased application of these synthetic organic polymers in the treatment and purification of water serves to increase their variety. This continuing development of new synthetic organic high molecular weight flocculants is a direct result of the need to remove the harmful products of technology.

This book delineates the basic types of high molecular weight, water-soluble organic compounds employed in the treatment and purification of water and/or in the treatment of industrial suspensions and sludges. Emphasis is given to high molecular weight organic compounds with flocculanting effects.

The authors have tried to present a clear review of known types of synthetic, natural, and/or chemically modified organic flocculants. Table 1 systematically divides these organic flocculants into basic groups according to their chemical composition as follows:

    I. Flocculants from natural sources (natural and chemically modified)
       A. Starch, modified starch, and starch derivatives
       B. Cellulose derivatives
       C. Other polysaccharides
       D. Vegetable gums and other vegetable substances, chemically modified
       E. Proteins and proteinaceous products

    II. Flocculants based on acrylic acid and methacrylic acid polymers
       A. Acrylic acid, its salts and copolymers
       B. Methacrylic acid and its derivatives
       C. Polyacrylamides, polymethacrylamides, their copolymers, derivatives and mixtures
       D. Polyacrylonitrile, its hydrolysis products, copolymers, etc.

    III. Flocculants based on polyvinyl alcohol and its derivatives
       A. Polyvinyl alcohol, its derivatives and copolymers
       B. Polyvinyl acetate and its derivatives
       C. Flocculants based on polyvinylamine, polyvinylsulfonic acid, polyacrolein

    IV. Flocculants based on maleic acid and its anhydride

    V. Flocculants based on polyamines and related substances
       A. Polyamines
       B. Condensation products of polyamines (and amines) and halogen hydrocarbons
       C. Various modified polymers—polyamino
       D. Polyamines and halohydrins
       E. Reaction products of amines (poly) and aldehydes

    VI. Flocculants based on polymers of alkylenimines (aziridines), ethylene (alkylene) oxides

    VII. Flocculants based on polyamides

    VIII. Flocculants based on polystyrene

    IX. Heterocyclic polymers (polymers of heterocyclic-containing or -forming monomers)

    X. Polymers of diallyldialkylammonium (and diallylammonium) salts

    XI. Polysulfones and polysulfines

    XII. Mixture of flocculants (synergistic combination)

About 25 years have elapsed since the early laboratory samples of organic flocculants were developed and tested. During this time span, the chemical industry has developed numbers of organic flocculants with increased and improved flocculating efficiency. At present there is a great variety of different types of organic flocculants available for use in the world market. They are sold under various trade names, none of which in any way impart information as to their composition or use. The greater part (about 80%) of available organic flocculants are synthetic; natural organic flocculants represent about one-fifth of the total.

Investigation of the sales of chemicals used in the United States for the treatment and purification of water and for the improvement of sludge characteristics before final water treatment, indicates that organic polymer consumption will increase at a greater rate than inorganic coagulant consumption [Gross, A.C.: *Environ. Sci. & Technol.*, 8, 414 (1974)] (see Figure 1). This higher consumption can be expected in spite of the fact that the unit price of organic flocculants is much higher than that of inorganic coagulants. The price of natural organic flocculants is in the range of $0.10 to $0.50/lb, while that of synthetic organic flocculants is in the range of $0.40 to $2.50/lb.

FIGURE 1: SALES OF WATER AND WASTEWATER TREATMENT CHEMICALS IN THE UNITED STATES

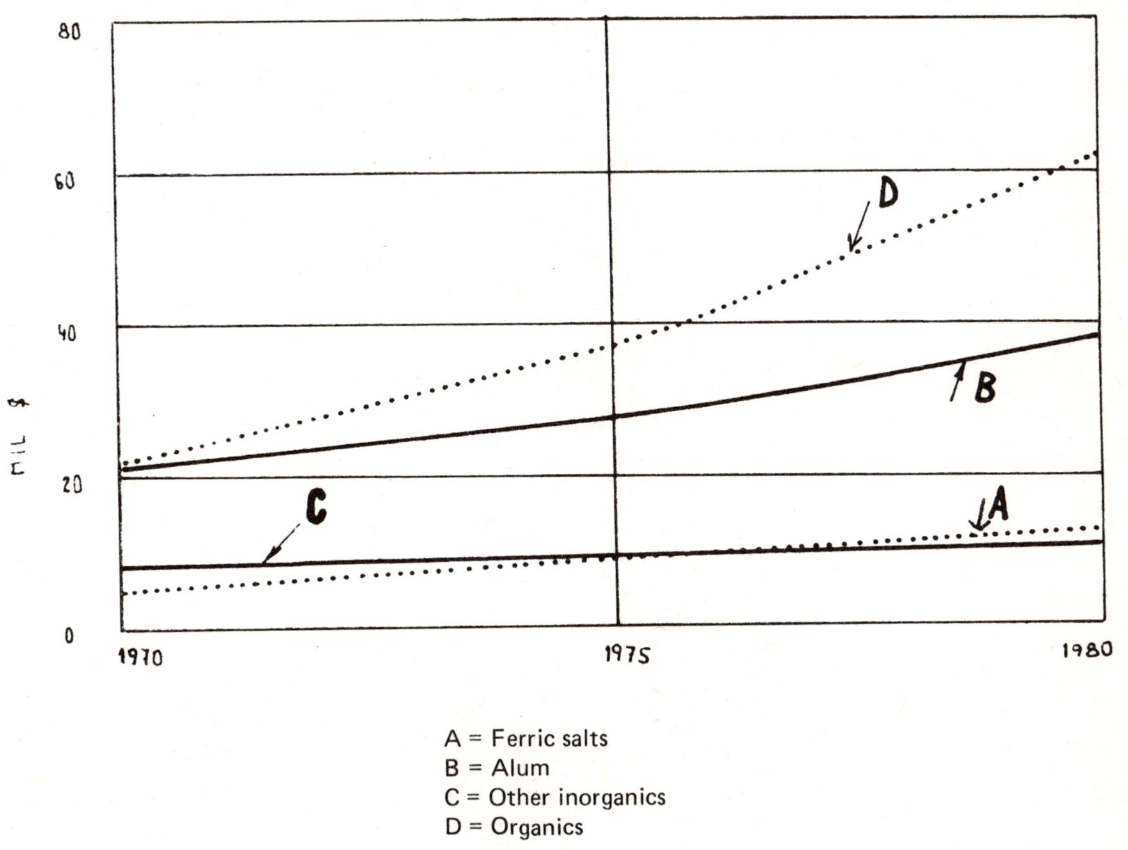

A = Ferric salts
B = Alum
C = Other inorganics
D = Organics

Source: Gross, A.C.; *Environmental Science & Technology*, 8, 414 (1974)

Tables 2 and 3 present a review of the basic characteristics of organic flocculants listed alphabetically by trade name. Table 2 gives detailed information regarding chemical composition and properties of these flocculants, while Table 3 describes their application (mostly in water treatment). Organic polymers used as emulsion breakers, scale inhibitors, and for turbulent drag reduction are not included. Table 4 gives a listing of manufacturers and/or distributors.

Table   1

Organic Flocculants
Classified by Chemical Composition

I. FLOCCULANTS FROM NATURAL SOURCES (NATURAL AND CHEMICALLY MODIFIED)

A. Starch, Modified Starch, and Starch Derivatives

Chemical Composition	Characteristic Group, and/or Outlined Preparation	Type A = Anionic C = Cationic N = Nonionic	Literature References	Examples of Application
1. Starch		N	Czech Pat. 141,711 (1971) /CA 77; 77031h/ Czech Pat. 121,886 (1967) /CA 69; 9889g/	Flocculation of colloidal phosphate ore in the presence of clay/U.S. Pat. 2,660,303 (1951)/; In bauxite processing/French Pat. 1,470,568 (1967), CA 67(18) 1967, 84102d/; Conditioning of sewage sludge in combination with inorganic salts /Brit. Pat. 1,188,394 (1970), CA 73; 7004v/; Sedimentation of coal sludges /Czech Pat. 141,711 (1971)/; Stabilizing drilling muds in deep wells/Czech Pat. 121,886 CA 69, 3889g/
2. Starch dispersions	By heating to 110°–150°C under pressure or treating aqueous solution of NaOH		German Pat. 2,158,130 (1973)/CA 79 (1973), 80919a/	Flocculants for red mud
3. Causticized starch			Jour. Soc. Chem. Ind., Transactions 1941, Vol. LX, p. 84	Clarifying of ore processing effluents/U.S. Pat. 3,414,512 (1968); CA 70 (1969), 49029a/; Coagulant aid/Environ. Health (India) 7, (1), 39 (1965); CA 66(14), 1967, 58778p/
4. Phosphoric acid derivates of starch		A	Brit. Pat. 1,309,473 (1973) /CA 79 (1973), 116739/; Japan 70-18, 217 (1970) /CA 74, 43746t/	As flocculant of tailings from coal flotation; As coagulants
5. Starch ether products - starch derivatives			Dutch Pat. 67-14, 110 (1969) /CA 71, 33239d/	Thickening dredged sludge; Improving of magnetic thickening,

Chemical Composition	Characteristic Group, and/or Outlined Preparation	Type A = Anionic C = Cationic N = Nonionic	Literature References	Examples of Application
			U.S. Pat. 3,697,420 (1972) /CA 78 (1973), 19988k/	retention aid in papermaking /ATIP 1970, 24(2), 61; 67830w/; Flocculation and accelerated sedimentation of solids in sewage (with iron ore weighting agent) /U.S. Pat 3,423,312 (1969); CA 70, 80713t/
a) Aminoethylethers of starch		C	Nippon Kagaku Kaishi 1973, (2), 233/CA 78 (1973), 126007m/	Flocculant of kaolin clay slurries, Flocculation of cellulose fiber slurries
b) Quarternary pyridinium (or ammonium) salt - starch ether derivatives; and/or gelatinized starch complex	e.g. N-(3-chloro-2-hydroxypropyl)pyridinium chloride, N-(3-chloro-2-hydroxypropyl)-2,6-dimethylpyridine chloride, N-(3-chloro-2-hydroxypropyl)-2,4,6-trimethylpyridinium chloride, N-(3-chloro-2-hydroxypropyl)-2-methylpyridinium chloride;	C	U.S. Pat. 3,346,563 (1964) /CA 68, 14279p/; U.S. Pat. 3,669,955 (1972) /CA 77, 77032j/	
	Gelatinization and etherifying of starch with (chlorohydroxypropyl) trimethylammonium salts;	C	German Offen. 2,131,560 (1972)/CA 76, 155916n/	
	Produced by reacting starch aqueous suspension with secondary or tertiary amine reacted with bis(2-chloroethyl)ether, or 1,2-bis(2-chloroethoxy)-ethane, or 1,4-bis(chloromethyl)benzene, or 1,3-dibromo-2-propanol	C	U.S. Pat. 3,639,209 (1972)	Flocculant in papermaking
c) Starch etherification with reaction product of epichloro-	$CH_2{-}CH{-}CH_2{-}N^+(CH_3)_3Cl^-$ with epoxide O	C	French Pat. 1,493,421 (1967)/CA 69, 107757d/;	Dewatering of cellulose fiber slurries, treatment of paper pulps

Chemical Composition	Characteristic Group, and/or Outlined Preparation	Type A = Anionic C = Cationic N = Nonionic	Literature References	Examples of Application
hydrine and compounds of type Me_3N (e.g., glycidyltrimethylammonium chloride)			French Pat. 1,518,535 (1968) /CA 71, 40508n/; Belgian Pat. 626,712 (1963) /CA 60, 10910e/	
d) Propyltriethylammonium halides and their starch derivatives	e.g., (1,3-dihydroxypropyl)-triethylammonium chloride	C	German Offen. 2,055,046 /CA 77, 77030g/	
e) Reaction products of starch with N-lower alkyl methylol triazones	R = a lower alkyl group (methyl, ethyl, propyl, isopropyl, butyl etc.)	C	U.S. Pat. 3,157,594 (1964).	Flocculant aids for treating turbid water, for sewage treatment with ferric chloride and lime
6. Graft copolymers of starch	a) Starch-polyacrylamide graft copolymer		U.S. Pat. 3,561,933 (1971)	Flocculating agents by the purification of wastewaters in the process industries, the clarification of water for industrial and domestic use, filler-retention aids in paper manufacture, the processing of ores in the recovery of metals therefrom Flocculant of aqueous suspension of colloidal kaolin clay, coal dust, flocculation of powdered coal suspensions /Kogyo Kagaku Zasshi 1971, 74(3), 456; CA 75, 40110n/ Flocculation of kaolinite suspensions /Kogyo Kagaku Zasshi 1971, 74(1), 91; CA 74, 142949t/

Chemical Composition	Characteristic Group, and/or Outlined Preparation	Type A = Anionic C = Cationic N = Nonionic	Literature References	Examples of Application
	b) Saponified polyacrylate-grafted starches		U.S. Pat. 3,377,302 (1968) /CA 68, 106249g/	As drilling mud additives, hydraulic fluids and flocculants
	c) Acrylamide, acrylic acid and β-(methacryloyloxy)-ethyltrimethylammonium monomethyl sulfate		Stärke 1973, 25(3), 83 /CA 79 (1973), 7122f/	Flocculant for bauxite ore red mud suspensions
	d) 2-Hydroxy-3-methacryloyloxypropyltrimethylammonium chloride		U.S. Pat. 3,669,915 (1972) /CA 77, 77033k/; J. Appl. Polym. Sci. 1970, 14(10), 2601-9	Flocculant of SiO_2, settling aid
7. Starch xanthate - cationic polymer complex				Mercury removal from wastewater/ Environ. Sci. Technol. 1973, 7(7), 614-19/
8. (Powder) compositions of starch, and/or starch derivatives				
a) + pulver CaO	Ratio 40:60 by weight		Czech Pat. 140,730 (1971) /CA 76, 156969c/	Purification of effluents containing Al-salts, flocculation of ore slurries
b) + pulver or granulated NaOH			Jap. Pat. 72-19,523 (1972) /CA 78 (1973), 47629n/ Jap. pat. 72-20,601 (1972) /CA 77 (1972), 166486c/	Flocculant of aqueous slurries of $Fe(OH)_3$, $Mg(OH)_2$
c) + pulver KOH			Jap. Pat. 72-25,984 (1972) /CA 78 (1973), 60004t/	
d) + water glass			Polish Pat. 63,338 (1971) /CA 76, 142660h/	
e) + $CaCl_2$				Flocculating reagents/Brit. Pat. 633,553 (1948)/
f) mixtures of inorganic salts Na_2CO_3, borax, Mx				Purification of sugar-refining liquids /Brit. Pat. 661,577 (1948), French Pat. 1,007,216 (1948)/
g) Oxidized starch			Japan 69-14,959 (1969) /CA 71, 114405g (1969)/	Flocculant aid in the treatment of water by a flocculant cloud/Vod.

Chemical Composition	Characteristic Group, and/or Outlined Preparation	Type A = Anionic C = Cationic N = Nonionic	Literature References	Examples of Application
				Hosp. B 1971, 21(2), 43 Czech; CA 78 (1973), 151300j/
h) Starch phosphates and inorganic salts	e.g., Sodium salt of starch phosphate and sodium silicate		Japan Pat. 70-18,217 (1970) /CA 74, 43746t/	
i) Carbamoyl ethyl starch hydrolyzate (and inorganic salts)			Japan Pat. 25,981(63) (1961) /CA 1964, 10244b/	
j) Polymer-polysaccharide-caustic alkali compositions	e.g., NaOH, pregelatinized starch (potato) or sorghum and sodium polyacrylate and/or acrylamide-acrylate copolymer		U.S. Pat. 3,541,009 (1970) /CA 74, 23449v/	Flocculation and settling of sludges from aluminum ore processing, of suspensions of iron oxide and calcium oxide from steel mill slurry (a waste product from a blast furnace)
k) Mixtures of causticized starch and polyacrylamide		N		To purify residual waters from coal washeries/Rev. Minelor 1968, 19(11), 480; CA 70 (1969), 117904j/
l) + polyacrylic and/or polymethacrylic acid	Starch products: corn starch, potato starch, amylose, sorghum	A	U.S. Pat. 3,575,868 (1971) U.S. Pat. 3,397,953 (1968) French Pat. 1,470,568 (1967) /CA 67(18), 1967, 84102d/	Clarification of sodium aluminate solutions obtained by the digestion of aluminiferous ores with alkali solutions
m) Aqueous solution of inorganic salts, reaction product of urea-resin and modified oxidized starch (polysaccharide-resin coagulants)			U.S. Pat. 3,285,849 (1966) /CA 66(8), 1967, 30269a/	Coagulants for aqueous suspensions
n) Mixtures of hydrophilic colloid and sodium or potassium aluminate	Colloids e.g.: hydrolyzed starch, naturally occurring gums		U.S. Pat. 3,082,173 (1963)	Coagulant aid to alum treatment of turbid waters
9. Amylopectin		N	French Pat. 1,489,195 (1967) /CA 68, 88379t/	Flocculation of finely divided minerals

B. Cellulose Derivatives

Chemical Composition	Characteristic Group, and/or Outlined Preparation	Type A = Anionic C = Cationic N = Nonionic	Literature References	Examples of Application
10.			Neth. Appl. 67-14,110 (1969); Hung. Pat. 155,855 (1969) /CA 71 (1969), 15883e/	Clarification of dredge mining wastewaters, flocculation of acidic solutions and sewage
a) Sodium carboxymethylcellulose		A	Neth. Appl. 67-14,110 (1969); USSR Pat. 105,847 (1957) U.S. Pat. 3,066,095 (1962)	Stabilizing drilling muds in deep wells, clarification of water with inorganic coagulants, bentonite clay/U.S. Pat. 3,066,095 (1962)/; Flocculation and separation of precipitated phosphates from wastewaters/U.S. Pat. 3,617,569 (1970)
b) Carboxymethylhydroxyethylcellulose		A	Brit. Pat. 703,964 (1951)	
c) Hydroxyethylcellulose, carboxymethylcellulose, hydroxyethyl/carboxymethylcellulose			U.S. Pat. 3,338,828 (1967) Brit. Pat. 703,964 (1951)	Coagulant aids in the treatment of water and aqueous wastes with inorganic coagulants and electrostatically precipitated fly ash; Flocculation of solid microparticles suspended in the water/Jap. Pat. 18,831(63), Sept. 19, 1961/
d) Aminoethylethers of cellulose (and/or of starches - aminoethyl derivatives of polysaccharides)			Nippon Kagaku Kaishi 1973, (2), 233-9; CA 78 (1973), 126007m/	Flocculation of suspended particles of microcrystalline cellulose and kaolin
e) Cellulose-acrylamide graft copolymers	e.g., Acrylamide		Kyoto Kogei Seu Daiagaku Serigakubu Gakuyuton Hokaku 1973, 7(1), 122 Agnew. Makromol. Chem. 1971, 20, 47/CA 76, 73937g/	As flocculants

C. Other Polysaccharides

Chemical Composition	Characteristic Group, and/or Outlined Preparation	Type	Literature References	Examples of Application
11.			Neth. Appl. 6,703,738 (1967) /CA 68, 14281h/	

Chemical Composition	Characteristic Group, and/or Outlined Preparation	Type A = Anionic C = Cationic N = Nonionic	Literature References	Examples of Application
12. a) Polymers of alginic acid		A	French Pat. 2,133,266 (1972), 2,032,724 (1970) /CA 75, 52665p/ Brit. Pat. 1,045,599 (1966) /CA 66(4), 1967, 13979p/	Flocculating agents for clarifying water
b) Sodium polyalginate		A	Jap. Kokai 73-20,768 (1973)	Dewatering of inorganic sludges and suspensions (in combination with cationic flocculant), clarification of sugar juice/U.S. Pat. 3,567,512 (1971), CA 74, 127946w/, coagulation aid/Kogyo Yosui 60, 24(1963); Recovery of water-soluble high-molecular-weight materials from food industry wastes/Japan Pat. 71-40,183 (1971); CA 78 (1973), 20000p/
13. Pectin (methyl pectate)			U.S. Pat. 3,505,033 (1970)	Clarification of a sulfuric acid digestion solution, prepared from the digestion of a titaniferous iron ore or slag
14. Agar-agar				
15. Polysaccharides, xylan type				Precipitation of red muds/USSR Pat. 307,065 (1971); CA 75, 11321y/
16. Polymers of polysaccharides and N-containing resins (polysaccharide-resin coagulants)	Starches, glycogen, inulin, cellulose, chitin, pectin, gum, etc. and modified polysaccharides produced by substitution of at least 4% of the OH radicals in the polysaccharides with e.g., OCH_2OH, OR, OAc, OCH_2COOR etc. radicals (where R represents such	N,C	U.S. Pat. 3,285,849 (1966) /CA 66(8), 1967, 30269z/	Flocculating agents for minute particles suspended in water, for coal dust, or industrial wastewaters

Chemical Composition	Characteristic Group, and/or Outlined Preparation	Type A = Anionic C = Cationic N = Nonionic	Literature References	Examples of Application
	radicals as H, Na, K, NH_4 and alkyl, Ac = acetyl). The N-containing resins include the cationic, anionic and nonionic resins, which are made from urea, melamine or mixtures thereof reacted with formaldehyde. (Use with inorganic salts)			
17. Microbial polysaccharides	e.g., Cyclodextrins and their derivatives, mostly ethers (from starch - Bacillus macerans)	A	U.S. Pat. 3,426,011 (1969)	
	Gums, e.g., produced from sorghum grain flour by bacteria of the species Xanthomonas campestris, simply treating with aluminum sulfate	C	U.S. Pat. 3,342,732 (1972) /CA 67(24), 1967, 109877a/	Flocculant for kaolin clay
	Synthesized by cultivating one or more of the microorganisms, e.g., Cryptococcus laurentii in a fermentation medium	C	U.S. Pat. 3,346,463 (1967) U.S. Pat. 3,406,114 (1968)	Flocculating agents for finely divided solids suspended in aqueous media, flocculation of slimes or finely divided solids in hydrometallurgical leach liques (e.g., potash ore, uranium ore)
	From yeasts (e.g., Pullularia pullulans)		U.S. Pat. 3,320,136 (1967)	
	A Xanthomonas microbial gum	C	U.S. Pat. 3,598,730 (1971)	Cationic flocculant

D. Vegetable Gums and Other Vegetable Substances, Chemically Modified

Chemical Composition	Characteristic Group, and/or Outlined Preparation	Type	Literature References	Examples of Application
18.			U.S. Pat. 3,515,666 (1970)	Clarification of water
a) Polygalactomannan gum (guar gum) and its cationic derivatives, mostly amino- and quaternary ammonium de-		C		Treatment of sewage or industrial effluents in combination with a polyvalent metal cation or cationic derivatives/U.S. Pat.

Chemical Composition	Characteristic Group, and/or Outlined Preparation	Type A = Anionic C = Cationic N = Nonionic	Literature References	Examples of Application
rivatives, made by reacting of gum with reactive amino compounds or reactive quaternary ammonium compounds				3,498,912 (1970); CA, 103518r/
b) Aminoethyl ethers of polygalactomannan gums (guar gum)		C	U.S. Pat. 3,303,184 (1967) /CA 67(22), 1967, 101246y/	For flocculating ore slimes, improving paper strength
c) Polyhydroxy derivatives of polygalactomannan	e.g., Guar gum with borax	C	U.S. Pat. 3,578,588 (1971) /CA 75, 38164t/	Coagulants for cellulosic compounds
19. Aminoethyl ethers of tamarind gum			U.S. Pat. 3,303,184 (1967) /CA 67(22), 1967, 101246y/	For flocculating ore slimes, improving paper strength
20. Aminoethylation of corn flour	e.g., Ethylenimine	C	German Offen. 2,218,805 (1972)/CA 78 (1973), 31774b/	
21. Locust beam gum with borax		C	U.S. Pat. 3,578,588 (1971)	Flocculation of cellulose fiber slurries
22. Nirmali fruits (Strychnos potatorum)				Clarification of coal washing effluents/Environ. Health 10(4), 265 (1968); CA 70 (1969), 108991h/
23. Reaction product of peeled fruits of Tamarindus indica, or Phaseolus mungo (beans), or Phaseolus radiatus (beans) with 35% NaOH			Indian Pat. 117,025 (1970) /CA 76, 49645x/ Indian Pat. 116,996 (1970) /CA 76, 49644n/ Indian Pat. 116,953 (1970) /CA 76, 49643v/	Flocculation of aqueous suspensions, sewage and industrial wastewaters
24. Mixture of ground carob seeds, Na$_2$SO$_4$ and H$_2$O, and/or Al and Fe salts			German. Offen. 1,297,048 (1969)/CA 71, 1969, 105113h/	Flocculant aid for potable water
25. Lignin derivatives	Compounds prepared by	A	U.S. Pat. 3,470,148 (1969)	As coagulation aids

Chemical Composition	Characteristic Group, and/or Outlined Preparation	Type A = Anionic C = Cationic N = Nonionic	Literature References	Examples of Application
	reacting lignin-containing materials (particularly kraft and sulfite lignins) and alkaline extracts of barks of trees, with compounds such as cyanuric chloride, 2,4-dichloro-6-methoxy-s-triazine, hydrazides, tetrakis-hydroxy-methyl phosphonium chloride; anionic character of the originally phenolic materials may be diminished by chemical modification, e.g., with an aldehyde, a triazine derivative		U.S. Pat. 3,600,308 (1971)	
26. Tannin (-based)			Brit. Pat. 1,310,491 (1973) /CA 78 (1973), 1513541/	Treatment of sewage sludge
27. Modified resin pitch		C		Purification of wastewaters containing emulsions, anionic surface-active substances, anionic sol. organic substances, dye wastewaters /Senryo To Jakuhi 1971, 16(12), 455; CA 77, 24507m/
28. Apple tree, cherry tree and peach tree resins			Tr. Nauchnoisled. Inst. vodosnab. Kanaliz. Sanit. Tech. 1970, 5, 153/CA 79 (1973), 107936t/	As flocculant
29. Wax dispersions formed by the emulsification of a natural or synthetic wax using a cationic, anionic or nonionic organic emulsifying agent	Examples of suitable waxes: naturally occuring animal and vegetable waxes (beeswax, spermacetti), earth waxes (ozocenite, ceresin),		U.S. Pat. 3,537,990 (1970)	As a flocculant in the clarification of water and industrial waste-waters and sewages

Chemical Composition	Characteristic Group, and/or Outlined Preparation	Type A = Anionic C = Cationic N = Nonionic	Literature References	Examples of Application
	paraffin wax etc. Examples of emulsifying agents: cationic—hydrochloride or acetate quaternary salts of C_{10}-C_{20} alkyl imidazolines; anionic—the ammonium and alkali metal salts of of C_{10}-C_{30} primary and secondary alkyl sulfonates and sulfates; nonionic—the C_7-C_{30} alkylphenoxy polyethyleneoxyethanols			
30. Chitin derivatives	A partially deaminated chitin and a partially deaminated and deacetylated derivative of chitin (chitosan)		U.S. Pat. 3,533,940 (1970)	As a coagulant to treat the impurity in an aqueous medium
31. Neutral soyabean milk				Coagulant for aqueous kaolin suspension/Kagaku Kogaku 29(5), 327-30 (1965); CA 67, 85325r/

E. Proteins and Proteinaceous Products

Chemical Composition	Characteristic Group, and/or Outlined Preparation	Type	Literature References	Examples of Application
32. Gelatin, glue		N	U.S. Pat. 3,658,184 (1972) /CA 77, 37035g/	Filter aid, removal of turbidity (U.S. Pat. 3,562,154 (1971); CA 74 (1971), 89152z/
33. Crosslinked collagen	Collagen derived organic nitrogenous colloidal substances of the protein class include glue, gelatin, water-soluble extracts from collagen-like materials, treated with crosslinking agents (e.g., for-	N	U.S. Pat. 2,937,143 (1960)	Flocculants in various industrial operations and clarification of natural turbid waters

Chemical Composition	Characteristic Group, and/or Outlined Preparation	Type A = Anionic C = Cationic N = Nonionic	Literature References	Examples of Application
	maldehyde, acetaldehyde, glyoxal and their homologs, methyl glyoxal and its homologs etc.) under different conditions			

II. FLOCCULANTS BASED ON ACRYLIC ACID AND METHACRYLIC ACID POLYMERS

A. Acrylic Acid, Its Salts and Copolymers

Chemical Composition	Characteristic Group, and/or Outlined Preparation	Type	Literature References	Examples of Application
34. Polyacrylates, mainly Na (NH$_4$, K)		A	French Pat. 2,133,134 (1972) /CA 79 (1973), 6109j/ French Pat. 1,571,371 (1969) /CA 72, 91788q/ German Offen. 1,961,099 (1970)/CA 73, 67331j/ Brit. Pat. 862,719 (1969) U.S. Pat. 3,657,378 (1972) /CA 77, 35232n/	Flocculation of solid particles in alkaline aqueous suspensions/ French Pat. 1,571,371 (1969)/ flocculant and thickener/German Offen. 1,961,099 (1970)/ Removal of clay from water streams, treating tar sands, recovery of water for recycle back into the hot water process/U.S. Pat. 3,487,003 (1969)/, flocculation of mineral suspensions and ore slurries/U.S. Pat. 3,418,237 (1968)/, separation of red mud from alumina/U.S. Pat. 3,445,178 (1969)/, French Pat. 1,470,568 (1967), Brit. Pat. 1,260,206 (1972), Jap. Kokai 72, 21392 (1972)/CA 79 (1973), 21143v/
	Poly(ammonium acrylate)	N	U.S. Pat. 3,658,771 (1972) /CA 77, 35240v/ USSR Pat. 204,569 (1967) /CA 69, 19785f/	
	Derivatives of acrylic and methacrylic acids, prepared by alkalinization of imino-esters of these acids		German East Pat. 61,950 (1968)/CA 70 (1969), 48038/ Japan Pat. 71, 10,955 (1971) /CA 77, 35285p/ Brit. Pat. 1,260,206 (1972) /CA 76, 141523d/	Processing finely suspended (waste) materials (coal fines, tailings etc.) /Czech Pat. 118,568 (1966); CA 20 (1967), 88519x/, flocculation of a suspension of precipitated calcium carbonate/Kogyo Kagaku Zasshi 69, 1195-6 (1966); CA 66, 116328e/, flocculation of positive charge suspension (Jap. Kokai 69, 18731) (1969)/, flocculation of phosphate mine wastes

Chemical Composition	Characteristic Group, and/or Outlined Preparation	Type A = Anionic C = Cationic N = Nonionic	Literature References	Examples of Application
				and coal washing waters effluents /U.S. Pat. 3,492,225 (1970); /CA 72, 82785z/, clarification of water together with the use of bentonite /U.S. Pat. 3,276,998 (1966)/, concentration of mineral suspensions through filtration and centrifugation/Chem. Tech. 1971, 23(4–5); 249; /CA 75, 38737g/, pressure filtration of mineral suspensions (kaolin)/Erzmetall 1961, 22(11), 524; /CA 72, 34459h/, water treatment with small-grained sand/Brit. Pat. 1,081,096 (1967)/, inhibiting incrustations in cooling towers and heat exchanger/German Offen. 1,808,824 (1969); /CA 71, 64003b/, flocculating oil- and clay-containing slimes/U.S. Pat. 3,723,310 (1973); /CA 79 (1973), 55450h/, improving of filterability of precipitated metal hydroxides, esp. $Mg(OH)_2$ (German East 57,116 (1967)/, filtration of mineral suspensions /Keram. Z. 1971, 23(1), 17–21; /CA 74, 143752k/, waste treatment for by-products/German Offen. 1,942,472 (1970); /CA 72, 136191g/, purification of wastewater from the polymerization of styrene/German Offen. 2,057,743 (1972); /CA 77, 52199f/, cleaning the wastewater of the cellulose, paper and paste industries/German Offen. 1,226,949 (1966); /CA 66(4), 1967, 13928w/, dewatering organic sludges by centrifugation/Brit. Pat. 1,225,635 (1971); /CA 77 (1972),

Chemical Composition	Characteristic Group, and/or Outlined Preparation	Type A = Anionic C = Cationic N = Nonionic	Literature References	Examples of Application
				143667y/ Dewatering of highly concentrated inorganic suspensions (with cationic polyelectrolyte)/Jap. Kokai 73, 20,768 (1973); /CA 79 (1973), 34909z/
35. Acryl acid copolymers				
a) Acrylic tetrakis (allyloxy) ethane copolymers	e.g., Acrylic acid-1,1,2,2-tetrakis (allyloxy) ethane copolymer		German Offen. 2,214,945 (1972)/CA 78 (1973), 16997b/	Thickening of aqueous suspensions
b) Acrylonitrile		A	German Offen. 2,057,743 (1972) Jap. Kokai 73, 20,768 (1973)/CA 79 (1973), 34909k/	Flocculation of wastewaters from pearl styrene production, dewatering of highly concentrated inorganic suspensions (in combination with cationic polyelectrolytes
c) Copolymers of acrylic acid and N-vinyl heterocyclic monomers	Prepared in a halogenated hydrocarbon medium	A,C	U.S. Pat. 3,284,414 (1966) /CA 66(6), 1967, 19060d/	As flocculants
d) Acrylic acid-acrylamide (and/or N-subst. acrylamide) copolymers		N,A	German Offen. 2,059,828 (1971)/CA 75, 12188z/ U.S. Pat. 3,374,143 (1968) /CA 68, 98488s/ Brit. Pat. 1,216,105 (1970) /CA 74, 64702r/	Purifying wastewaters containing proteinaceous materials, white water recovery (vacuum or flotation systems), flocculation of montmorillonite suspension, flocculation of mineral suspensions and mineral slurries/U.S. Pat. 3,418,237 (1968), U.S. Pat. 3,425,802 (1969)/
e) Copolymer sodium ethylene sulfonate-sodium acrylate		A		Flocculation of kaolin and precipitated calcium carbonate suspensions /Kogyo Kagaku Zasshi 69(7), 1335, 1339 (1966); /CA 66, 116331a, 116332b/
f) Acrylic terpolymers	Prepared, e.g., of acrylamide, partially neutralized acryl-acid and acrylonitrile (50-90:6-35:4-15%)	A	German Offen. 2,025,725 (1970)/CA 74, 54794h/	Flocculant and coagulant aid

Chemical Composition	Characteristic Group, and/or Outlined Preparation	Type A = Anionic C = Cationic N = Nonionic	Literature References	Examples of Application
36. Mixtures of:				
a) Acrylate polymer with cupric or nickel ions				Settling of solids suspended in water, flocculation process for copper sulfide ore/U.S. Pat. 3,524,811 (1970); /CA 73, 80387b/
b) $Al_2(SO_4)_3$, polyacrylic acid ester and sodium alginate			Japan Pat. 72, 26,584 (1972) /CA 78 (1973), 33700s/	
c) Polymers of acrylic acid and acrylic acid amide		A		Flocculants for water treatment, sewage purification and dewatering of sludge in metal-processing and metal-working industries /Wasser, Luft, Betr. 1969, 13(4), 129; /CA 71, 42041k/
37. Copolymers of 2-sodiosulfo-ethyl acrylate and acrylamide		A	U.S. Pat. 3,312,671 (1967) /CA 67, 12074y/	As flocculants
38. N-alkyl-substituted aminoalkyl esters of acrylic acids		C	U.S. Pat. 3,171,805 (1965)	Removal of clay from water streams, treating tar sands, recovery of water for recycle back into the hot water process/U.S. Pat. 3,487,003 (1969)/, clarification of surface waters/U.S. Pat. 3,483,120 (1969)/, clarification of turbid water simultaneously by treatment with alkali permanganate
a) Poly(diethylaminoethyl-methylacrylate)		C	French Pat. 1,570,656 (1969)/CA 72, 45618z (1970)/	Flocculating agents for titanium sulfate liquors
b) Poly(diethylaminoethyl-acrylate)acetate		C	French Pat. 1,570,656 (1969)/CA 72, 45618z (1970)/	Flocculation of clay in water
c) Poly(dimethylaminoethyl-acrylate)		C	U.S. Pat. 3,014,896 (1961)	Flocculation of wastewaters, removal of anionic detergents
			U.S. Pat. 3,171,805 (1965)	Sludge filtration
d) Dimethylaminopropylacryl-ate		C	U.S. Pat. 3,171,805 (1965)	Flocculation of wastewaters, removal of anionic detergents, sludge filtration

B. Methacrylic Acid and Its Derivatives

Chemical Composition	Characteristic Group, and/or Outlined Preparation	Type A = Anionic C = Cationic N = Nonionic	Literature References	Examples of Application
39. Polymethacrylates (Na, K, NH$_4$)		A	French Pat. 1,571,371 (1969)	Flocculation of coal suspensions, clay ores; clarification of pulps /USSR Pat. 226,582 (1968); CA 70, 39298j/, pressure filtration of kaolin suspensions (Sklar. Keram. 1973, 23(1), 13; /CA 79 (1973), 117891j/ Flocculation and an improvement in the settling of mineral water suspensions or ore pulps/U.S. Pat. 3,418,237 (1968)/
	Ammonium polymethacrylate	A	German Offen. 2,036,581 (1972)/CA 77, 20669t/ USSR Pat. 196,004 (1967) /CA 68, 52199v/	Clarification of slurries from coal beneficiation plants/Koks Chim. 1970, (12), 4; 1972, (5), 10/, /Izv. Dnepropetrovsk. Gorn. Inst. 1971, 57, 124; /CA 78, (1973), 128156h/ Removal of clay from water streams, treating tar sands, recovery of water for recycle back into the hot water process/U.S. Pat. 3,487,003 (1969)/ Thickener, flocculant and settling aid /German Offen. 2,036,581 (1972)/ Inhibiting incrustations in cooling towers and heat exchangers /German Offen. 1,808,824 (1969); CA 71, 64003b/
40. Methacryl acid-methacrylamide copolymers				Flocculation of phosphate mine waters, coal wash water effluents /U.S. Pat. 3,492,225 (1970); CA 72, 82785h/
41. Hydrolyzed acrylate-vinyl alcohol copolymer	e.g., The acrylic monomer component: the lower alkyl esters of acrylic and methacrylic acid (methacrylate, methylacrylate,		U.S. Pat. 3,442,799 (1969) S. African 67,01,885 (1968) /CA 70 (1969), 89887k/	Clarification of raw sewage, settling red mud from suspensions of bauxite in caustic solutions

Chemical Composition	Characteristic Group, and/or Outlined Preparation	Type A = Anionic C = Cationic N = Nonionic	Literature References	Examples of Application
	acrylamide), alkali metal or ammonium salts of methacrylic or acrylic compounds etc. The vinyl ester component: esters of lower alkanols (C_1 to C_4) capable of being hydrolyzed e.g., Sodium acrylate/vinyl alcohol copolymer prepared as heterohydrolysis product of a methylacrylate/vinylacetate copolymer			
42. Compounds of methacrylic acid (and/or alkali metal or alkaline earth metal salts) and cyclized dimethacryloylimide (mol. ratio of 1:1)		A		Clarification of raw sugar juice/U.S. Pat. 3,033,782 (1962)/
43. N-alkyl-substituted aminoalkyl esters of methacrylic acid		A		Removal of clay from water streams, treating tar sands, recovery of water for recycle back into the hot water process/U.S. Pat. 3,487,003 (1969)/
a) Dimethylaminoethylmethacrylate		C		Flocculation of precipitated ferric hydroxide from molybdic acid solutions/USSR Pat. 338,549 (1972); /CA 77, 105423h/
b) Polymers prepared by amination of poly-(β-chloroethylmethacrylate)			USSR Pat. 319,611 (1971) /CA 76,100659j/	
c) Methacrylic β-aminoalkyl esters, prepared by alkylation	e.g., Poly(β-dimethylaminoethylmethacrylate) or poly[1,3-bis(dimethylamino)isopropyl-methacrylate] with CH_3I, $PhCH_2Cl$ or EtBr		Vysokomol. Soedin. Ser. A, 1971, 13(9), 2139;/CA 76, 14988j/	Flocculation of clay suspensions

Chemical Composition	Characteristic Group, and/or Outlined Preparation	Type A = Anion C = Cation N = Nonionic		
d) Aminopropargyl esters of methacrylic acid (quaternary polymers)	Prepared by radiation polymerization			
44. Reaction products of a polyamine and an acrylate type compound (amino-amido polymers)	Prepared, e.g., from methylacrylate and ethylene diamine	C	/CA 73, 122152d/	…emulsions/U.S. Pat. 3,509,047 (1970)/, corrosion inhibitors/U.S. Pat. 3,514,250 (1970)/
45. Polymers of (methacryloyloxyethyl) trimethyl ammonium	Polymers of the type: $MeN^+CH_2CH_2OOCHC{=}CH_2$ and/or $CH_2{=}CMeCOOCH_2CH_2N^+HMe_2$ and their copolymers with methacrylic acid, methacrylamide, N-vinylpyrrolidone	C	German Offen. 1,935,476 (1971)/CA 74, 79326v/	Flocculation of industrial wastewaters containing Ca or heavy metal ion hydroxides Flocculants for metal hydroxides
46. Polymethacryloxyethyl sulfonium halide		C	U.S. Pat. 3,269,991 (1966) U.S. Pat. 3,280,081 (1966) U.S. Pat. 3,207,656 (1965) U.S. Pat. 3,332,835 (1967)	
47. Polymers from poly(glycidyl methacrylate) (with amines)	e.g., Poly(glycidylmethacrylate) and triethylenetetramine	C	Jap. Kokai 72-29,482 (1972) /CA 79 (1973), 79733d/ Nippon Kagaku Kaishi 1972, (1), 184-8/CA 76, 149922h/ 1973,(6), 1201-5/CA 79 (1973) 79953a/	

C. Polyacrylamides, Polymethacrylamides, Their Copolymers, Derivatives and Mixtures

Chemical Composition	Characteristic Group, and/or Outlined Preparation	Type		
48. Polyacrylamides Polyacrylamide-type flocculating agents		N	USSR Pat. 150,795 (1962) USSR Pat. 123,482 (1959) USSR Pat. 123,483 (1959) Hung. Pat. 155,163 (1968) /CA 70 (1969), 48280p/	Flocculation and settling of ore pulps and mineral suspensions/U.S. Pat. 3,418,237 (1968), U.S. Pat. 3,480,761 (1969)/, in the process of manufacturing alum/U.S. Pat.

Chemical Composition	Characteristic Group, and/or Outlined Preparation	Type A = Anionic C = Cationic N = Nonionic	Literature References	Examples of Application
	5% aqueous solution of acrylamide containing N,N'-methylenebisacrylamide (0.00923 % vol)		Polish Pat. 56,009 (1968) /CA 70 (1969), 97733x/ French Pat. 1,466,796 (1967)/CA 67(16), 1967, 740025s/ Belgian Pat 589,293 (1960) Brit. Pat. 1,106,573 (1968) /CA 68, 95376y/ Brit. Pat. 1,139,917 (1968) /CA 70 (1969), 58458s/ Jap. Pat. 69,01,426 (1969) /CA 71, 125674v/ German Offen. 2,143,549 (1973) German Offen. 2,125,502 (1972)/CA 78 (1973) 59426g/ German Offen. 1,961,101 (1970)/CA 73, 78253m/ German Offen. 2,122,415 (1972) Przem. Chem. 1972, 5(1), 743 U.S. Pat. 3,332,922 (1972) /CA 67(20), 1967, 91246g/ Ambio 1972, 1(5), 180 /CA 78 (1973), 61956y/	3,425,802 (1969)/ methyl derivative /Tr. Inst. Met. Obogash. Akad. Nauk Kaz. SSR 1971, No. 41, 48; /CA 76, 101781e/; Slimes elimination in potash ore treatment/U.S. Pat. 3,451,788 (1969); Vesti Adak. Nauk Belorusk, SSR, Ser. Chim. Nauk 1967, (2), 111; 1967, (3), 75; Chim. Prom. (Moscow) 1972, 48(9), 681/; Flotation of ore manganese products /Obogash. Rud. 1973, 18(1), 16; /CA 79 (1973), 56228d/; Centrifugal-flocculation process for purifying the recycle waters concentrating mills reprocessing manganese ores/Vodosnabzh.-Sanit. Tech. 1970(8), 12(Russ.); /CA 74, 67446w/; Flocculants for nonferrous ore dressing, e.g., Cu-ore (copper concentrates) /Rudy Metale Niezelaz. 1970, 15(10), 549;/CA 75, 24090d/, purification of Cu-mine waters/Kem. Ind. 1970, 19(9), 443/; Thickening of slurries at beneficiation Au-ores containing 30% limonite /Obogashch. Met. Polez. Iskop, 1970, 23; /CA 75, 15399y/; The recovery of nickel catalysts /U.S. Pat. 3,640,897 (1972), /CA 76, 90702b/; Recycles waters at magnetic ore dressing plants/Sb. Nauch. Tr., Nauch.-Issled. Projkt. Inst. Obogashch. Aglomet. Rud. Chern. Met. 1972, No. 13, 15; /CA 79 (2973), 45607a/; Flocculation and filtration of kaolin
	Polymerization by use of a redox catalyst system with a constant free-radical concentration; anionic flocculant alk. hydrolyzed			

Chemical Composition	Characteristic Group, and/or Outlined Preparation	Type A = Anionic C = Cationic N = Nonionic	Literature References	Examples of Application
				and coal suspensions/Kogyo Kagaku Zasshi 96(6), 1199, (1969), /CA 66, 116329f/; 69(6), 1187 (1966), /CA 66, 116326c/; Erzmetall 1969, 22(11), 524, /CA 72, 34459h (1970)/; Flocculation of flotation waste rock /Ugol 1971, 46(10), 57; /CA 76, 27785r/; Purification of industrial effluents during processing of placer deposit /Kolyma 1970, No. 3, 38, /CA 76, 37171b/; Coagulation of wastewater, slurries from yellow phosphorus production in continuous thickener settlers/Khim. Prom. (Moscow) 1972, 48(1), 857 (Russ.); /CA 78, (1973), 140108b/; Ultrasonic aggregation of mineral suspensions/Tr. Inst. Met. Obogashch. Akrad. Nauk. Kaz. SSR, 1968, 32, 53, /CA 71, 52456p/; Flocculation of precipitated calcium carbonate suspensions/Kogyo Kagaku Zashi 96(6), 1199 (1969)/; Purification of recirculating waters of the fosforit plant concentrating mill /Sanit. Tech. 1967, 7–12 (Russ.), /CA 71, 128426 (1969)/; Treating paramagnetic slurries/Ind. Wastes 1969 (Nov.) IW/5–IW/9—polyacrylamide type/; Decantation and filtration of industrial suspensions/French Pat. 1,492,037 (1967), /CA 68, 115205v/; Clarification of neutralized wash

Chemical Composition	Characteristic Group, and/or Outlined Preparation	Type A = Anionic C = Cationic N = Nonionic	Literature References	Examples of Application
				waters from the machine industry /Probl. Isopolz. Okhr. Vod. Resur. 1972, 124–8 (Russ.), /CA 77 (1972), 168399a/; Purification of wastewaters from mining enterprises/Vodosnabzh. Sanit. Tekh. 1970, (11), 12 (Russ.), /CA 74, 90938s/; Purification of coal mine waters /Vodosnabzh, Sanit. Tekh. 1973, (2), 8–10 (Russ.), /CA 79 (1973), 45373w/; Flocculation of coal sludges and slime slurries/Paliva 46, 312 (1966), /CA 67(20), 1067, 92620mm/; /Vuglishta 1970, 25(1), 21, /CA 73, 37207f; Tr. Inst. Gornych, Iskop., Moscow 1968, 24(2), 66, /CA 70 (1969), 89442m/; Dewatering of fine-grained fractions of coal, clayey muds and flotation tailings/Polish Pat. 67,075 (1973), /CA 79 (1973), 147765t/; Treating of wastewater from zinc and lead mines and processing plants/ Gas, Woda Sanit. 1970, 44(9), 308, /CA 74, 90937r/; Precipitates of titanium and magnesium production wastewaters /Vodosnabzh. Sanit. Tekh. 1973, (4), 11, /CA 79 (1973), 23336j/; Purification of mine waters used in dust removal in mining processes /Nauch. Tr., Karagand. Territ. Otd. Vost. Nauch.-Issled. Inst. 1971, No. 1,307, /CA 78 (1973), 33619x (Russ.)/;

Chemical Composition	Characteristic Group, and/or Outlined Preparation	Type A = Anionic C = Cationic N = Nonionic	Literature References	Examples of Application
				Clarification of water from open-hearth furnace gas purifications in radial settling tanks/Vodosnabzh. Sanit. Tekh. 1972, (2), 16, /CA 76, 131209s/;
				Precipitation of suspended matter in electric furnace gas purifier off-gases, i.e., wastewaters/Vodosnabzh. Kanaliz., Gidrotekh. Sooruzheniya 1969, No. 11, 8, /CA 73, 112696v/;
				Dust from Wagner-type converters /Kogyo Josui 1968, (115), 45, /CA 69, 38534q/;
				Flocculation of Fe oxide dust from the OG-type (steel) converter in aqueous solutions/Kogyo Josui 1968, (122), 32, /CA 74, 15581s/;
				Purification of wastewaters from metallurgical plants/Pure Appl. Chem. 1972, 29(1-3), 323, /CA 77, 92513k/;
				Wastewaters in metal mining industries/Vop. Technol. Orab. Vady, Prom. Pitevogo Vodosnabzh. 1969, No. 1, 77, /CA 75, 91117f/; Nauch. Tr., Tashkent. Univ. 1971, No. 403, 108/;
				Chemical purification of wastewaters from degreasing facilities from pickling departments/Ochistka Povtornoe Izploz. Stochnykh Vod Urale 1967, 2, 40-9, /CA 71, 84371y/;
				Purification of wastewaters from the scrubbing of metallurgical assemblies /Vodosnabzh. Sanit. Tekh. 1971, (3), 9, /CA 75, 24962q/; Liteinoe Proizvod. 1972, (5), 28, /CA 78

Chemical Composition	Characteristic Group, and/or Outlined Preparation	Type A = Anionic C = Cationic N = Nonionic	Literature References	Examples of Application
				(1973), 33583f/; German Offen. 2,154,462 (1972), /CA 77, 79317e/; Removing petroleum and petroleum products from wastewaters/Sb. Tr., Mosc. Inzh.-Stavit. Inst. 1971, No. 87, 1193, /CA 78 (1973), 151231n/; Purification of washwaters from service stations operating on ethylated gasoline/Ochistka Stochnykh Prir. Vod. 1970, 99, /CA 76, 6494e/; Purification of oil-containing wastewaters/Ochistka Stochnykh Prir. Vod. 1970, 123, /CA 76, 6485c/; Purification of wastewaters containing oil emulsions by pressure flotation/Vodosnabzh, Sanit. Tekh. 1972, (1), 16, /CA 76, 89772t/; Flocculating agent in drilling fluids for rotary drilling of oil wells/U.S. Pat. 3,637,031 (1972), /CA 76, 155109k/; Purification of wastewaters by removal of benzene, ethylbenzene and styrene/Rev. Chim. (Bucharest) 1971, 22(12), 732, /CA 76, 144571d/; Clarification of sugar juice/U.S. Pat. 3,567,512 (1971), /CA 74, 127946w/, /Fr. Demand 2,143,201 (1973)-anionic, /CA 79 (1973), 7130g/; Clarification of conveyor-wash waters for sugar beet/Kharkchova Prom. 1971, (6), 29, /CA 76, 131179g/; Removing starch from sweet sorghum juices/South African 68, 08,553 (1969), /CA 72, 134420g/; Treatment of the wastewater from

Chemical Composition	Characteristic Group, and/or Outlined Preparation	Type A = Anionic C = Cationic N = Nonionic	Literature References	Examples of Application
				the fisheries industry/Hokusuishi Geppo 1973, 30(3), 22, /CA 79 (1973), 96581y/; Coagulation of lignin in waste sulfite pulp/Jap. Kokai 73, 54,201 (1973), /CA 79 (1973), 127534p/; Treatment of pulp mill effluents /Kami Pa Gikyoshi 1973, 27(6), 283, /CA 79 (1973), 55122c; Tr. Nauch.-Issled. Inst. Osm. Khim. (USSR), 21, 102 (1970)/; Paper mill wastes/Eisei Kagaku 1972, 18(4), 274, /CA 78 (1973), 7539b; Shizuoka-Ken Seishi Kogyo Shikensho Shiken Kenkyu Hokokusko 1972, (24), 105, /CA 79 (1973), 23332c/; Pulp Paper Mag. Can. 68(8), T 367 (1967), /CA 67(16), 1967, 74604e/; Kobunshi Kako 1971, 20(1), 660, /CA 76, 158009p/; Bum. Prom. 1972, (5), 18, /CA 77, 38825a/; German Offen. 1,961,101 (1970), /CA 73, 78253m/; Purification of tannery wastewaters /Kozh.-Obuv. Prom. 1972, 14(5), 44, /CA 78 (1973), 33631v/; 1971, 13(10), 32, /CA 76, 6501e/; Purification of wastewaters from scouring and dyeing plants/German Offen. 2,107,120 (1971), /CA 75, 132783r/; Removal and recovery of mercury and copper from wastewaters, flocculation of precipitated Hg and Cu from wastewaters/German Offen. 2,122,415 (1972), /CA 78 (1973), 33675n/;

Chemical Composition	Characteristic Group, and/or Outlined Preparation	Type A = Anionic C = Cationic N = Nonionic	Literature References	Examples of Application
				Dehydration of sediments from viscose fiber production sewage on vacuum filters/Sanit. Tekn. Vol. XXIII Nauk. Konf. Leningrad Inz.-Strit. Inst., Sb., 1965, /CA 65, (1966), 3551g/;
				Clarification and purification of surface and industrial sewage containing turbid matter/German Offen. 1,271,641 (1968), /CA 69, 61420h/;
				Conditioning and dewatering sewage sludge from clarifying equipment /German Offen. 2,103,970 (1972), /CA 78 (1973), 19983c/;
				Treatment of sewage sludge/Brit. Pat. 1,310,491 (1973), /CA 78 (1973), 151354e/; U.S. Pat. 3,642,619 (1972), /CA 76, 158068g/; French Pat. 1,168,959 (1958);
				Chemical conditioning of sewage sludges/Brit. Pat. 1,191,632 (1970), /CA 73, 18347k/;
				Flocculation of activated sludge Bum. Prom. 1969, (8), 10, /CA 71, 92838w (1969)/;
				Precipitation process for the removal of phosphates from wastewaters /U.S. Pat. 3,617,569 (1971), /CA 76, 27845k/;
				Dewatering of wastewater deposits /Bum. Prom. 1972, (2), 14, /CA 76 (1972), 117275d/;
				Sewage sludge treatment/South African 68, 08,074 (1969), /CA 72, 47213y (1970)/;
				Dewatering of sludge from converters gas purifiers in metallurgical plants (drum vacuum filters)/Vodosnabzh.

Chemical Composition	Characteristic Group, and/or Outlined Preparation	Type A = Anionic C = Cationic N = Nonionic	Literature References	Examples of Application
				Sanit. Tekh. 1969, (8), 14, /CA 72, 24352d/;
				Treatment of water and sludge (coal fines)/Aust. Chem. Eng. 1969, 10(5), 19, /CA 71, 73801n/;
				Conditoning of sludge/Eisel Kagaku 8, 1, 23 (1972), /CA 77 (1972), 38852g/;
				Thickening of tailings from the magnetic concentration plants /Obogashch. Polez. Iskop, 1971, No. 9, 81 (Russ.), /CA 77 (1972), 142638c/;
				Dewatering of organic sludges by centrifugation/Brit. Pat. 1,225,635 (1971), /CA 77 (1972), 143667y/;
				Flocculation of clay water suspensions/U.S. Pat. 3,487,003 (1969), U.S. Pat. 3,516,932 (1970), /CA 73, 69703f/; U.S. Pat. 3,637,491 (1972), /CA 76, 158197y/;
				Filtration of mineral suspensions /Keram. Z 1971, 23(1), 17, /CA 74, 143752k/;
				Filter aid by removing of softening products from softened water/U.S. Pat. 3,171,803 (1965)/;
				Filter softener by turbid water filtration/U.S. Pat. 3,171,801 (1965)/;
				Purification of river water/Tr. Bakinsk. Fil. Vses. Nauch.-Issled. Inst. Vodosnabzh. Kanaliz., Gidrotekh. Sooruzh, Inzh. Gidrogeol. 1970, No. 5, 7–13 (Russ.), /CA 76, 49763j/, /Energetik 1969, 17(3), 22–4 (Russ.), /CA 71, 33292z/;
				As flocculant (with alum) in coagula-

Chemical Composition	Characteristic Group, and/or Outlined Preparation	Type A = Anionic C = Cationic N = Nonionic	Literature References	Examples of Application
				tion of colored water/Energetik 14(4), 14 (1966) (Russ.), /CA 65(4), 1966, 5218d/; As filter aid in the purification of potable water/Nauch. Tr. Akad. Kommunal. Khoz. 1969, No. 52, 46–52 (Russ.), /CA 74, 45451s/; coagulant and filter aid in rapid sand filtration/Kogyo Josui 1966, No. 90, 35 (Japan), /CA 69, 80075r/; Removal of alluvium in cooling water systems/U.S. Pat. 3,288,640 (1966), /CA 66(8), 1967, 29803n/;
49. Poly(acrylamide sulfate)			German Offen. 2,054,647 (1970) /CA 76, 34724g/	
50. Polygalactomannan-acrylate adduct			U.S. Pat. 3,346,555 (1967) /CA 68, 90222e/	Flocculation of bentonite or kaolin suspensions
51. Poly(methacrylamide)		N		Flocculation of clay in water/U.S. Pat. 3,487,503 (1969)/ Clarification of red mud pulps/USSR Pat. 285,912 (1970), /CA 75, 8038n/;
52. Acrylamide-acrylic acid (salts) copolymers	Storage-stable granules (mixture with starch)	A	German Offen. 2,143,549 (1973) /CA 79 (1973), 19699n/ U.S. Pat 3,658,772 (1972) Indian Pat. 123,577 (1971) /CA 77, 20996y/ Jap. Kokai 73, 01,082 (1973) /CA 78 (1973), 125388z/ German Offen. 2,112,231 (1971) /CA 76, 60323m/ U.S. Pat. 3,658,772 (1972) /CA 77, 35478d/	Clarifying surface waters using poly-electrolytes and sand/U.S. Pat. 3,350,302 (1967); Clarification of sugar juice (settling and vacuum filtration steps)/U.S. Pat. 3,508,965 (1970), /CA 73, 57421z/; Clarification of raw sugar juice/Proc. Queens. Soc. Sugar Cane Technol. 1973, 40, 243, /CA 79 (1973), 68047g/; Clarification of the digested slurry in the process of manufacturing alum

Chemical Composition	Characteristic Group, and/or Outlined Preparation	Type A = Anionic C = Cationic N = Nonionic	Literature References	Examples of Application
				/U.S. Pat. 3,425,802 (1969)/; Flocculation and settling solids in mining and other industrial operations, e.g., copper ore slurries, coal wash waters/U.S. Pat. 3,479,282 (1969); 3,479,284 (1969)/; Clarifying aqueous suspensions of iron oxide particles in a steelmaking process/U.S. Pat. 3,549,527 (1970), /CA 74, 57155m/; Flocculation of coal slime/U.S. Pat. 3,717,574 (1973), /CA 78(19), 1973, 151363g/; Removal of soluble phosphates from sewage effluent/U.S. Pat. 3,617,542, /CA 76, 27842g/; Removal of soluble phosphates from sewage/U.S. Pat. 3,506,570 (1970); German Offen. 2,016,768 (1970), /CA 74, 79316s/; Reducing the amount of lime needed to remove color from paper mill wastewaters/U.S. Pat. 3,578,587 (1971), /CA 75, 38164t/; Treatment of paper pulps/French Pat. 1,518,535 (1968), /CA 71, 40508n/; Flocculating agent and settling aid for removing suspended solids from raw influent sewage slurry/U.S. Pat. 3,479,283 (1969), /CA 72, 47207z (1970) German Offen. 1,961,101 (1970), /CA 73, 78253m/;
			Brit. Pat. 1,259,306 (1972) French Pat. 1,452,107 (1966)/CA 66(16), 1967, 67070d/ French Demande 2,140,116 (1973)/CA 79, 1973,	Filtration of sludge through cloth with the addition of wood flour/Brit. Pat. 1,259,306 (1972)/; Flocculation of C, SiO_2, $Al(OH)_3$ dispersions in rolling mills wastewaters/French Pat. 1,452,107

Chemical Composition	Characteristic Group, and/or Outlined Preparation	Type A = Anionic C = Cationic N = Nonionic	Literature References	Examples of Application
			54056x/ German Offen. 1,961,101 (1970)/CA 73, 78253m/ German Offen. 2,064,101 (1972)/CA 77 (1972), 127438n/	(1966),/CA 66(16), 1967, 67070d/; Flocculation and thickening of sludges/Terres Eaux 21(55), 1968, 27; German Offen. 1,961,101 (1970)/CA 73, 78253m/; Settling aids
53. (Meth)acrylamide-(meth)-acrylic acid salts copolymers		A		Clarification of and phosphates removal from sewage/U.S. Pat. 3,506,570 (1970)/CA 73, 7000r; U.S. Pat. 3,617,569 (1970)/; Filtration of asbestos cement suspensions/Z. Prikl. Khim. (Leningrad), 1968, 41(12), 2630p (Russ)/CA 70, (1969), 69089y/
54. Aqueous suspension of acrylamide, copolymer of acrylamide-acrylic acid (9:1) and Na polyphosphate (and/or Na hexametaphosphate, Na tripolyphosphate)			Jap. Pat. 71, 24,881 (1971) /CA 77, 21020m/	Beneficiation of sedimentation of suspension particles, e.g., Mg(OH)$_2$
55. Copolymers of acrylamide and a sulfoalkyl acrylate salt		A		Flocculation and settling of inorganic particles in salt solutions; In refinement of water-soluble ores /U.S. Pat. 3,617,573 (1971)/CA 76, 27835g/
a) Copolymer of 2-sodiosulfo-ethylacrylate and acrylamide		A	U.S. Pat. 3,312,671 (1967) /CA 67, 12074y/	
b) Copolymer of acrylamide and sodium sulfopropyl acrylate (85–98:2–15), -sodiosulfoethyl acrylate, sodiosulfobutyl acrylate, and/or potassium salts	Mostly acrylamide-Na sulfo-propylacrylate			Flocculation of inorganic particles (e.g., clay) from aqueous solutions /South African 71, 05,480 (1972); /CA 78 (1973), 86697c/

Chemical Composition	Characteristic Group, and/or Outlined Preparation	Type A = Anionic C = Cationic N = Nonionic	Literature References	Examples of Application
56. Copolymer containing acrylamide (60–65%), acrylic or methacrylic acid (10–25%), acrylic or methacrylic acid esters (10–25%), acrylonitrile (0–20%)			French Pat. 2,108,635 (1972)/CA 78 (1973), 31734g/	In paper manufacture
57. Copolymer of acrylamide and water-soluble acid esters (e.g., methyl acid maleate)		C	U.S. Pat. 3,201,304 (1965)	Flocculation of cellulose fiber slurries
58. Acrylamide-maleic anhydride-methacrylic acid copolymer (e.g., 170:1:80 by weight)			French Pat. 1,518,535 (1968)/CA 71, 40508n/	Dewatering of cellulose fiber slurries
59. Acrylamide-/(dimethylamino)-methyl/acrylamide terpolymers	e.g., Copolymer of acrylamide-methyl methacrylate /(dimethylamino)methyl /acrylamide		German Offen. 2,249,925 (1971)/CA 79 (1973), 79480u/; U.S. Pat. 3,658,772 (1972) /CA 77, 35478d/	Flocculating and retention aid in papermaking
60. Copolymers of acrylamide and quaternary ammonium compounds		C		
a) Copolymers of acrylamide and diallyl methyl (β-propionamido)ammonium halide (chloride)	Ratio 50:50 or 75:25%			Sewage flocculation and sludge filtration, solids-liquid separation processes/U.S. Pat. 3,514,398 (1970)/
b) Acrylamide (β-methacryloyloxyethyl) trimethylammonium methyl sulfate		C	U.S. Pat. 3,480,541 (1969) /CA 72, 1970, 24402v/ Japan Kokai 73, 38288 (1973)/CA 79 (1973), 106330d/	Flocculation and sedimentation of solids of raw influent sewage slurry /U.S. Pat. 3,480,541 (1969), /CA 72, 24402v (1970)/; Emulsion breaking agents; Treating digested diluted sewage slurry (elutriation steps)/U.S. Pat. 3,414,513 (1968)/; Dewatering a concentrated sewage

Chemical Composition	Characteristic Group, and/or Outlined Preparation	Type A = Anionic C = Cationic N = Nonionic	Literature References	Examples of Application
				slurry (vacuum filtration)/U.S. Pat. 3,414,514 (1968), /CA 70 (1969), 50328d/; U.S. Pat. 3,472,767 (1969)/; In sugar industry (settling or vacuum filtration steps)/U.S. Pat. 3,479,221 (.1969)/
c) Copolymers of acrylamide and diallylamine (or diallyl-ammonium) compounds	e.g., Copolymer acrylamide-diallyl (β-carbamoethyl)-ammonium chloride			Flocculation and coagulation of wastewaters/Neth. Appl. 6,611,834 (1967); /CA 67(16), 1967, 76156j/
d) Quaternaries: acrylamido-propylbenzyl dimethylam-monium hydroxides, diallyl-dimethylammonium chlo-ride, etc.		C	U.S. Pat. 3,171,805 (1965)	Sewage treatment (settling or filtra-tion steps sludge filtration), removal of anionic detergents
e) Copolymers of acrylamide-N-[2-dimethylamino)ethyl] -3-[(2-(dimethylaminoethyl)-amino] -2-methylpropion-amide-N-[2-(dimethylamino-ethyl] -methacrylamide	Mol. ratio 80:1-15:19–4		U.S. Pat. 3,661,868 (1972) /CA 77, 49419x/	Flocculation and settling aid of raw sewage, dewatering of sewage sludge
61. Copolymers of acrylamido sul-fonic acids (alkali metal or ammonium salts) (sulfonate polymers)			U.S. Pat. 3,692,673 (1972) /CA 78 (1973), 20009y/	As coagulant aid in conjunction with inorganic coagulants
a) Acrylamide-vinylaromatic sulfonates copolymers			U.S. Pat. 2,909,508 (1956)	
b) Acrylamide-sodium vinyl sulfonate copolymer	Ratio e.g., (85-98:15-2%)	A	U.S. Pat. 3,617,572 (1971) /CA 76, 47879c/	Flocculation of kaolin suspension and precipitated calcium carbonate /Kogyo Kagaku Zasshi 69, (7), 1339; /CA 66, 116332b/; 69, (7), 1335 (1966), /CA 66, 116331a/; In the refinement of the water-soluble ores (e.g., separation of clay from a borax ores solution, brines);

Chemical Composition	Characteristic Group, and/or Outlined Preparation	Type A = Anionic C = Cationic N = Nonionic	Literature References	Examples of Application
				Flocculation and settling of inorganic particles from salt solutions in ore treatment/U.S. Pat. 3,617,572 (1971)/
62. a) Vinyl lactam-acrylamide (or methacrylamide) copolymer (and/or hydrolyzed)	N-vinyl-2-pyrrolidone-acrylamide, N-vinyl-caprolactam-acrylamide, N-vinyl-5-methyl-2-pyrrolidone-acrylamide	C	U.S. Pat. 3,412,020 (1968) /CA 70, 39064n/ French Pat. 1,479,951 (1967)/CA 67(24), 10967, 109184x/	Flocculation of paper mill sludge, clarification of water for industrial use, mine underground water, settling aid for ore and coal fines/U.S. Pat. 3,412,020 (1968)/
b) Quaternized vinyl lactam-acrylamide copolymers	Prepared by treating of acrylamide-vinyl lactam copolymers (25–95:5–75% by weight) with HCHO and Me$_2$NH in presence of MeCl or PhCH$_2$Cl		German Offen. 2,116,921 (1971)/CA 76, 73195w/	Flocculating agents (used in papermaking)
63. Polyacrylamide terpolymers: copolymers of acrylamide and vinylpyridine, or vinylacetate, or vinylstyrene, or vinylethers, or vinylhalides, or isobutylene	e.g., Acrylamide-4-vinyl-N-benzyl-pyridinium chloride (C)	A,C	U.S. Pat. 3,478,003 (1969) /CA 72, 1970, 22402w/ Sin. Vysokomol. Soedin. 1972, 70:7/CA 79 (1973), 19567t/	Settling of ore pulps and mineral suspensions Flocculation of silicagel suspensions
64. Copolymers of methacrylamide and compounds of type: $CH_2{=}CRCONH(CH_2)_nCHCOOH$ $\quad\quad\quad\quad\quad\quad NH_2$ where R = H or Me, n = 2–4	e.g., Copolymerization of α-amino acids (or salts) and acrylamide (or methacrylamide) in ratio (15–85):(85–15%); from removal of Ca^{++} ion copolymer of N-ϵ-acryloyl-L-lysine and acrylamide (1:1)		Jap. Kokai 73, 23,690 (1973) /CA 79 (1973) 23392z/	Elimination of heavy metal ions from wastewaters (Cd, Co, Cu, Ni, Mn, Fe)
65. Reaction product of glyoxal and copolymer of acrylamide-acrylic acid			Jap. Kokai 73-12,289 (1973) /CA 79 (1973), 96664z/	Purification of industrial effluents (e.g., papermaking wastewaters)
66. Copolymer of acrylamide (carboxylated) and epichlorohydrin			Jap. Kokai 73-12,894 (1973) /CA 79 (1973), 96665a/	Purification of wastewaters

Chemical Composition	Characteristic Group, and/or Outlined Preparation	Type A = Anionic C = Cationic N = Nonionic	Literature References	Examples of Application
67. Copolymers of itaconic acid (or salts), acrylamide and acrylic acid or methacrylic acid (94:3–5:2% by weight)			German Offen. 2,207,795 (1972)/CA 77 (1972), 165531b/	Treatment of water
68. Polyacrylamide containing colloidal silicic acid			Jap. Pat. 71,21,735 (1971) /CA 76, 4491c/	As flocculant
69. Chelate forming amino acid addition polymers	Prepared by condensation of compound containing acrylic groups and amino acid containing $\geqslant 2$ active H ions, e.g., copolymer of glycin-N,N-methylenebis-acrylamide		German Offen. 2,258,535 (1973)/CA 79 (1973), 67075c/	Removal of heavy metals from waste-waters (e.g., Cu)
70. Copolymer of diacetone acrylamide (2.5–65% by weight) and various comonomers (e.g., substituted acrylamide, diallyl ammonium compounds)		C	U.S. Pat. 3,489,681 (1970) /CA 72, 82781d/	Clarifying of water having solids suspended, green liquor clarification in papermaking, flocculant and coagulant aid
71. Alkoxylated polyacrylamides	e.g., Modified by propylene oxide		German Offen. 2,021,006 (1971)/CA 76, 86780h/	Flocculation of bentonite suspensions, aqueous or acidic suspensions of cellulose
72. Product of acrylamide and formaldehyde			USSR Pat. 150,795 (1962)	
73. Acetaldehyde modified acrylamide			Kyoto Kogei Seri i Daigaku Seriigakubu Gakujutsu Hokoku 1973, 7(1), 122 /CA 79 (1973), 53986p/	
74. Carbamoyl polymers	Prepared by reaction of polyacrylamide, HCHO Me$_2$NH, with piperidine, morpholine, methylaminoethanol or diethanolamine	C	U.S. Pat. 3,539,535 (1970) /CA 74, 13588a/	

Chemical Composition	Characteristic Group, and/or Outlined Preparation	Type A = Anionic C = Cationic N = Nonionic	Literature References	Examples of Application
a) Polyacrylamide modified with paraformaldehyde and amines (e.g., reaction product of polyacrylamide)			German Offen. 2,159,028 (1973)/CA 79 (1973), 67515w/	Flocculation of enzyme in *Bacillus subtilis* cultures
b) Reaction product of poly-acrylamide and sec. amines and formaldehyde	Mixture of [(dimethylamino)-methyl] acrylamide-41%, methylolacrylamide-46%, acrylamide-13%	C	German Offen. 2,249,602 (1973)/CA 78 (1973), 160478a/ German Offen. 2,163,246 (1972)/CA 78 (1973), 85290c/	Flocculation of ore slurries, e.g., ilmenite
c) N-substituted-(N'-dialkyl-aminoalkyl)acrylamides	e.g., N-(diethylaminoethyl)-acrylamide, N-(diethyl-aminoethyl)methacryl-amide, N-(dipropylamino-ethyl)acrylamide, N-(piper-idylmethyl)acrylamide	C	U.S. Pat. 3,171,805 (1965) German Offen. 2,206,564 (1972)/CA 77 (1972), 165533d/	Flocculation of wastewaters, removal of anionic detergents, sludge filtration
d) Aminoalkyl-acrylamides		C		Removal of clay from water streams /U.S. Pat. 3,487,003 (1969)/
e) Polyamino-N-alkyl-acryl-amides		C		Clarification of turbid waters/U.S. Pat. 3,483,120 (1969)/
f) Acrylamide-(diethylamino)-ethyl acrylate methosulfate	As acrylic copolymer pearls		German Offen. 2,043,663 (1972)/CA 77, 20395a/	Flocculant for activated sewage
g) Acrylamide-diethylamino-ethyl methyl acrylate-co-polymers of acrylamide-dimethylaminoethyl meth-acrylate copolymer			French Pat. 1,570,656 (1969)/CA 72 (1970), 45618k/ French Pat. 1,518,535 (1968)/CA 71, 40508n/	Flocculating agent for titanium sulfate liquors, treatment of paper pulps
h) N-[(dimethylamino)methyl]-polyacrylamide	Prepared from polyacryl-amide (1% hydrolyzed amide groups) parafor-maldehyde and dimethyl-ammonium oxalate		German Offen. 1,442,398 (1973)/CA 79 (1973), 19719u/	Flocculation of starch suspension

Chemical Composition	Characteristic Group, and/or Outlined Preparation	Type A = Anionic C = Cationic N = Nonionic	Literature References	Examples of Application
75. Copolymers of the type:	$$\left[\left(CH_2-\underset{\underset{x}{\overset{\displaystyle C=O}{\mid}}{\overset{\displaystyle R}{\mid}}C\right)\left(CH_2-\underset{y}{\overset{\displaystyle R}{\mid}}C-C\right)_a\right]_n$$ where x is a cationic grouping, y is a nonionic solubilizing radical, a indicates the ratio of nonionic to cationic units, n denotes the degree of polymerization The cationic groupings: $-O-CH_2-CH_2-N\big<^R_R$ $-O-CH_2CHOHCH_2N\big<^R_R$ $-NHCH_2CH_2N\big<^R_R$ $-NHCH_2CH_2CH_2N\big<^R_R$ R = methyl, ethyl, hydrogen Nonionic groupings: $-CONH_2,\ -CO_2CH_2CH_2OCH_3,\ -OH$	C	German Offen. 2,136,887 (1973)/CA 78 (1973), 125190d/ German Offen. 2,134,717 (1972)/CA 76, 141727y/ U.S. Pat. 3,692,673 (1972)	Flocculating of negatively charged colloids/U.S. Pat. 3,014,896 (1961)/ Retention aid in papermaking/German Offen. 2,136,887 (1973)/
76. Condensation product of polyacrylamide and polyamine	e.g., Polyacrylamide, 3-(dimethylamino)propylamine	C	U.S. Pat. 3,503,946 (1970) /CA 72, 122369j/	Flocculant for sewage sludge, for ilmenite ore slimes, paper mill

Chemical Composition	Characteristic Group, and/or Outlined Preparation	Type A = Anionic C = Cationic N = Nonionic	Literature References	Examples of Application
(water-soluble)	and ethylene glycol			white water, effluent from mining operations
77. Crosslinked polyacrylamide	e.g., Prepared by subjecting a water-soluble polyacrylamide to transamidation with water-soluble poly-(primary amine) (e.g., ethylenediamine)	C	U.S. Pat. 3,488,720 (1970) /CA 72, 67746c/	Flocculant for solids in dilute aqueous suspension, sewage, coal slimes, paper mill white water
78. Cationic resins from acrylamide	Prepared e.g., from N-methacrylamide (or acrylamide), mono- or diethylamine, epichlorohydrin and tetraethylenepentamine (or pentaethylenehexamine)		Jap. Kokai 72, 42,998 (1972)/CA 78 (1973), 163754y/	Water clarification
79. Polymers of poly(methylmethacrylate)	Prepared e.g., by aminolysis of poly(methylmethacrylate) with polyethylenepolyamines	C	Nippon Kogaku Kaishi 1972, (1), 184/CA 76, 149922h/ U.S. Pat. 3,445,441 (1969)	Flocculating and oil emulsion breaking agents
80. Oxaminated polyacrylamides	e.g., Polyacrylamide modified 1-(diethylamino)-2,3-epoxypropane; 1-(dibutylamino)-2,3-epoxypropane; 1-(trimethylammonio)-2,3-epoxypropane chloride		German Offen. 2,053,675 (1972)/CA 77, 62984x/	Flocculating agents of bentonite suspensions
81. Copolymer poly[N-(2-aminoethyl)acrylamide] -epichlorohydrin	Prepared by hydrolysis of poly(vinylimidazoline) (obtained by reacting polyacrylonitrile with ethylene diamine) and treating of hydrolysis product with epichlorohydrin		U.S. Pat. 3,640,936 (1972) /CA 76, 129140n/	As flocculant and dewatering agent in papermaking

Chemical Composition	Characteristic Group, and/or Outlined Preparation	Type A = Anionic C = Cationic N = Nonionic	Literature References	Examples of Application
82. Graft copolymers of acrylamide methylamine-epichlorohydrin			U.S. Pat. 3,697,370 (1972) /CA 78 (1973), 59976m/	Flocculating agent for sewage
83. Polymers of polyethyleneimine and acrylamide-N-methylol-acrylamide copolymer		C	Jap. Pat. 72, 19,521 (1972) /CA 78 (1973), 98658d/	Flocculation of mine wastewaters
84. β-Morpholinopropionamide and acrylamide			German Offen. 2,136,384 (1973)/CA 78 (1973), 125154w/	As flocculant, retention aid and filter aid
85. Methylolated polyacrylamide			U.S. Pat. 3,747,676 (1973) /CA 79 (1973), 116801x/	Thickener for oil recovery
86. Sulfomethyl polyacrylamide	Prepared by treating HCHO and sulfurous acid salts		French Pat. 2,036,570 (1970)/CA 75, 141688r/ German East Pat. 71611 (1970)/CA 73, 67411h/ Jap. Kokai 73-59,664 (1973)/CA 79, (1973) 149227t/	Filtering aids for metal hydroxide precipitates Treatment of dye waste solution (in combination with high molecular weight condensation product of ethylenediamine and epichlorohydrin)
87. Copolymers polythioacrylamide or chelate polymers of polythioacrylamide			German Offen. 2,159,504 (1973)/CA 79 (1973), 57464q/	Removal of mercury salts from waste solutions
88. a) Synergistic mixtures of polyacrylamide (or polyacrylamide based flocculant) and cationic polymers	Cationic polymers: cationic hydroxy-containing polymeric esters of an ethylenically unsaturated acid [e.g., poly-2-hydroxy-3-(methacrylyloxy) propyl-trimethylammonium chloride]	C	U.S. Pat. 3,321,649 (1967) /CA 67(12), 1967, 55304e/	Separation of finely divided solids from aqueous suspensions, e.g., phosphate slime from fluorapatite ore
b) Mixture of water-soluble resinous amino condensation polymers and cationic vinyl addition polymer	e.g., A water-soluble homopolymer of acrylamide, in which about 35% of the amide groups were		U.S. Pat. 3,652,479 (1972) /CA 77, 6800x/ U.S. Pat. 3,409,546 (1969) /CA 70, 22757z/	Conditioning and dewatering of raw sewage sludge

Chemical Composition	Characteristic Group, and/or Outlined Preparation	Type A = Anionic C = Cationic N = Nonionic	Literature References	Examples of Application
	substituted on the nitrogen atom with dimethylaminomethyl moiety, and amino condensation resin (reaction product of a polyalkylene polyamine and ethylenedichloride)			
c) Mixtures of condensation product of urea (or guanidine-aldehyde) and polyacrylamide			German Offen. 1,243,643 (1967)/CA 67(22), 1967, 103160z/	Flocculation agent for aqueous suspensions
d) Mixture of modified polyacrylamide (6–35%), binding agent of inorg. salt (10–55%), and water (30–70%)-block form	e.g., Wastes from zinc-electroplating: anionic polyacrylamide-25%, Na_2CO_3 -70%, water 5%, and/or polyethyleneoxide-1%, water-2%, Na_2CO_3-97%		German Offen. 2,229,426 (1973)/CA 78(1973), 151379s/	As flocculant, e.g., for $Zn(OH)_2$ from metal finishing wastes from electroplating
e) Mixtures of polyacrylamides and partially hydrolyzed products			Jap. Pat. 72, 26,583 (1972) /CA 78 (1973), 33699y/	Flocculation of aqueous suspensions of $Mg(OH)_2$ (pH 10.6)
f) Mixture of polyacrylamide and hydrolyzed polyacrylamide (15% acrylic acid) in ratio of 3:1 and 1:3			Jap. Pat. 72, 25,077 (1972) /CA 78 (1973), 88411d/	Flocculation and removal of $\geqslant$10 g/l suspension matter
g) Synergetic compositions of nonionic polyacrylamide-base polymer and Fe^{3+} salt			U.S. Pat. 3,655,522 (1972) /CA 77, 38885v/	Removal of phosphates from wastewater
h) High molecular weight complex prepared, e.g., by treating of cationic poly(2-methyl-N-vinylimidazoline) with anionic polyacrylamide		A,C	Jap. Pat. 72, 19,522 (1972) /CA 78 (1973), 88413f/	Flocculants for wastewater treatment (settling aid, filtration aid, and clarification)

D. Polyacrylonitrile, Its Hydrolysis Products, Copolymers, etc.

Chemical Composition	Characteristic Group, and/or Outlined Preparation	Type A = Anionic C = Cationic N = Nonionic	Literature References	Examples of Application
89. Copolymers acrylonitrile-acrylic acid, -methacrylic acid, -maleic acid, and/or their salts, methacrylonitrile-vinylacetate, etc.		A		
90. Hydrolysis products of polyacrylonitrile (hydrolyzed polyacrylonitrile)	Hydrolysis products containing active groups $-CONH_2$ and COOH(Na) in various proportions depending on hydrolysis stage	N,A	God. Vissh. Minno-Geol. Inst., Sofia 1966–67 (Publ. 1968), 13, 117-31 (Bulg.)/CA 73, 112013v/; Sb. Tr., Gos. Nauk.-Issled. Inst. Cvent. Metal No. 19, 273-88(1962) Cvent. Metal 39(9), 63-4 (1966) (Russ.) /CA 66(16), 1967, 67992f/ Nauch. Tr., Tashkent. Gos. Univ. 1970, No. 399, 236–52 (Russ.)/CA 77 (1972), 165098n/	In processing of alum manufacturing /U.S. Pat. 3,425,802 (1969)/; Flocculation and sludge thickening /Terres Eaux 21(5), 1968, 27/; Flocculation in paper production /Sb. Tr. Ukr. Nauch.-Issled. Inst. Tsellyul. Bum. Prom. 1969, No. 12, 58–66; /CA 75, 7631p/; Flocculation and settling of ore pulps and mineral suspensions/U.S. Pat. 3,418,237 (1968)/; Flocculation of mineral and coal suspensions/Proc. Inst. Miner. Proces. Congr., 9th 1970, 444/CA 74, 78541z/; Filtration of asbestos-cement suspensions /Issled. Stroit. Mater. Izdeliyam. 1971, 54–61;/CA 79 (1973), 107729c/; Flocculation of red mud/Cvet. Metal. 39(9), 63–4 1966/
91. $HONH_2$-modified polyacrylonitrile			Jap. Kokai 73–59,663 (1973) /CA 79 (1973), 149226s/	Decolorizing agent for dye wastewaters
92. Polymer prepared by hydrolysis (90–100) acrylonitrile-vinylacetate-α-methylstyrene terpolymer with NaOH			Rom. Pat. 52,280 (1970) /CA 73, 121313v/	Water-clarifying flocculant
93. Compounds of acrylonitrile and amines	e.g., Polyacrylonitrile and N,N'-dimethyl-1,3-propropanediamine		German Offen. 2,056,032 (1971)/CA 75, 64865s/	Flocculating agent for purification of papermaking wastewaters, dewatering and filter aid of papermaking wastewater sludges

Chemical Composition	Characteristic Group, and/or Outlined Preparation	Type A = Anionic C = Cationic N = Nonionic	Literature References	Examples of Application
a) Flocculants prepared by reaction of polyacrylonitrile and compounds of the type $H_2NCHR_1(CHR_2)_nCH(OH)R_3$	Where R_1, R_2, R_3 are H or Me, n = 0 or 1, number of C-atom $<$4, e.g., polyacrylonitrile and ethanolamine	C	Jap. Pat. 72, 26,585 (1972) /CA 78 (1972), 88416j/	Flocculants for wastewater, filter aid for kaolin aqueous suspensions
b) Coagulant aid obtained by reacting a polyacrylonitrile with excess ethylene diamine				Filtration, centrifugation, sedimentation and decantation of aqueous suspensions of finely divided water insoluble solid materials, such as minerals or organic matter/U.S. Pat. 3,288,707/1966); /CA 66(10), 1967, 40595s/
94. Linear or crosslinked organo-silica polymers				
a) Copolymer acrylonitrile-silica (silicon dioxide)		A,C	U.S. Pat. 3,592,834 (1971) /CA 75, 110712q/	As dispersion and flocculation agent
b) Polyacrylonitrile and silicate-based polymer			Vzejmodejstvie Vodorastvorim Polielektrolit. Disperzn. Sist. 1970, 64-5 (Russ.)/CA 76 (1972), 154532f/	Purification of wastewater

III. FLOCCULANTS BASED ON POLYVINYL ALCOHOL AND ITS DERIVATIVES

A. Polyvinyl Alcohol, Its Derivatives and Copolymers

Chemical Composition	Characteristic Group, and/or Outlined Preparation	Type	Literature References	Examples of Application
95. Polyvinyl alcohol		N		Dewatering organic sludges by centrifugation/Brit. Pat. 1,225,635 (1971) /CA 77 (1972), 143667y/
96. Copolymers of vinyl alcohol (vinyl compounds) and acrylic (or maleic) acid, and/or salts and derivatives	Acrylic monomer components are the acrylic and methacrylic acid derivatives, such as methyl esters, amide or nitrile	A		Clarification agent for separating suspended solids from raw sewage/U.S. Pat. 3,442,799 (1969); /CA 71 (1969), 15877f/
97. Crotonic acid-vinyl alcohol copolymer			German Offen. 2,057,047 (1971) /CA 75, 110874u/	As flocculant for aqueous suspensions of alum and $Fe_2(SO_4)_3$

	Chemical Composition	Characteristic Group, and/or Outlined Preparation	Type A = Anionic C = Cationic N = Nonionic	Literature References	Examples of Application
98.	Vinyl amine-vinyl alcohol copolymers			U.S. Pat. 3,715,336 (1973) /CA 78 (1973), 125367s/	As flocculant
99.	Polyvinyl alcohol carbamates	e.g., Prepared by reaction of polyvinyl alcohol and urea		Jap. Pat. 18,827(63) (1961) /CA 60 (1964), 14256a/	
100.	Modified polyvinyl alcohol resins			Jap. Pat. 18,828(63) (1961) /CA 60 (1964), 14255e/	
101.	a) Thiuronium salts of polyvinyl alcohol b) Reaction product of polyvinyl alcohol, thiourea or an alkyl-substituted thiourea, and a strong mineral acid		C	Jap. Pat. 18,832(63) (1961) /CA 60 (1964), 14255f/ U.S. Pat. 3,148,142 (1964)	Flocculant and settling aid for kaollinite aqueous suspensions
102.	Polyvinyl alcohol-amino aldehyde reaction products (acetalizing polyvinyl alcohol)	e.g., Prepared by acetalizing polyvinyl alcohol with an aminoaldehyde (e.g., α,α-dimethyl-β-dimethyl-aminopropionaldehyde; α,α-dimethyl-β-(2-ethylhexyl-amino)propionaldehyde; β-aminobutylaldehyde	N	U.S. Pat. 3,166,497 (1965) U.S. Pat. 3,170,809 (1965)	Clarification agent for inorganic materials, particularly soil and clay (bentonite and kaollinite)

B. Polyvinyl Acetate and Its Derivatives

	Chemical Composition	Characteristic Group, and/or Outlined Preparation	Type	Literature References	Examples of Application
103.	Partially quaternized homopolymers of vinyl esters of α-halogenated aliphatic carboxylic acids or copolymers of vinyl esters of α-halogenated aliphatic carboxylic acids and other polymerizable monomers (e.g., vinyl acetate, methyl methacrylate, styrene) in which some of α-halo-carboxy groups are quaternized	The preferred ester: vinyl-chloroacetate the quaternizing agent: pyridine, triethylamine, bipyridyl and thiourea, vinyl acetate	C	Brit. Pat. 1,073,823 (1967) /CA 67(16), 1967, 76188w/ U.S. Pat. 3,432,430 (1969) Soc. Chem. Ind. (London) Monogr. No. 25, 131–40 /CA 68, 30531v/	Clarification of surface waters, removal of turbidity and color (10–30% quaternized), purification of sewage (30–60% quaternized)/U.S. Pat. 3,432,430 (1969)/

Chemical Composition	Characteristic Group, and/or Outlined Preparation	Type A = Anionic C = Cationic N = Nonionic	Literature References	Examples of Application
104. Copolymers of vinyl acetate:				
a) With maleic anhydride (also partially esterified) or with maleic acid, and/or its salts or ethers, or with acrylic acid (methacrylic acid), or with itaconic acid anhydride			U.S. Pat. 3,715,307 (1973) /CA 78 (1973), 101851u/	For conditioning of heat exchanger water
b) Crotonic acid-polyvinyl acetate copolymer				
c) Poly(Na acrylate)-maleic anhydride-vinyl acetate				Flocculation of phosphate mine waters, coal washing waters/U.S. Pat. 3,492,225 (1970)/CA 72, 82765h/
d) Copolymer of vinyl acetate with N-vinyl-succinimide and 2-dimethylaminoethyl-vinyl ether, or tert-Bu-N-vinylcarbamate			Jap. Pat. 72, 23,281 (1972) /CA 78 (1973), 85304k/	

C. Flocculants Based on Polyvinylamine, Polyvinylsulfonic Acid, Polyacrolein

Chemical Composition	Characteristic Group, and/or Outlined Preparation	Type A = Anionic C = Cationic N = Nonionic	Literature References	Examples of Application
105. Polyvinylamine				Flocculation and thickening of sludges /Terres Eau 21(55), 1968, 27-28; /CA 67 (1969), No. 4, 14242n/
106. Reaction products of amines and polyvinyl chloride, or copolymer of styrene-vinyl chloride, chlorinated paraffins (C_{20}–C_{30}), chlorinated polyolefins, copolymers of chlorodifluoroethylene-vinylidene fluoride	Amines, e.g., $H_2N(CH_2)_3NMe_2$, $BuNH_2$, Me_2NH or Me_3H		German Offen. 2,120,398 (1971)/CA 76, 73212k/	Flocculating agent, e.g., in papermaking
107. Polyacrolein	$CH_2{=}CH{-}CHO$	N	French Pat. 1,397,707 (1965) German Offen. 1,229,045 (1966)/CA 66 (10), 1967,	Sewage and industrial wastewater flocculation/German Offen. 1,299,566 (1962)/

Chemical Composition	Characteristic Group, and/or Outlined Preparation	Type A = Anionic C = Cationic N = Nonionic	Literature References	Examples of Application
			38666x/ Brit. Pat. 1,063,058 (1967) /CA 67, 14722g/	

IV. FLOCCULANTS BASED ON MALEIC ACID AND ITS ANHYDRIDE

Chemical Composition	Characteristic Group, and/or Outlined Preparation	Type	Literature References	Examples of Application
108. Maleic anhydride oligomers			German Offen. 2,159,172 (1972)/CA 77, 79464a/	Water treatment and boiler scale removal
109. Copolymers based on cross-linked maleic anhydride			U.S. Pat. 3,554,985 (1971) /CA 74, 91069w/	Concentration of viruses from sewage and excreta/Appl. Microbiol. 1959, 18(6), 1007/CA 72 (1970), 47154e/
a) Vinylacetate-maleic anhydride copolymer	Molar ratio 1:1, calcium-sodium salt	A		Clarification of raw sugar beet juice /U.S. Pat. 3,033,782 (1962)/; Corrosion inhibitors, industrial cooling water systems/U.S. Pat. 3,715,307 (1973)/CA 78 (1973), 101851u/
b) Reaction product of isobutylene-maleic anhydride copolymer and dimethyl-aminopropylamine (copolymers containing half-amides of olefinic anhydrides)		A	U.S. Pat. 3,157,595 (1964) /CA 64(1966), No. 9, 12977g/	Viruses concentration from water /JAWWA, Vol. 65 (1973), No. 3, p. 200/
c) Polymers of polyvinyl-methylether-maleic anhydride	Flocculants based on polymer having a linear hydrocarbon structure and containing in a side chain a hydrophilic group from the group consisting of carboxylic acid, carboxylic acid anhydride and carboxylic acid salt groups; Composition with colloidal clay and/or soda ash, sodium bicarbonate			Coagulation and settling agent for finely divided solids, employed in conjunction with colloidal clay (as bentonite, saponite)/U.S. Pat. 3,276,998 (1966)

Chemical Composition	Characteristic Group, and/or Outlined Preparation	Type A = Anionic C = Cationic N = Nonionic	Literature References	Examples of Application
d) Copolymers and terpolymers of olefinic anhydrides	e.g., Maleic anhydride, or isobutylene-maleic anhydride treated by diamines (e.g., dimethylamino-propylamine); Obtained products have 50% monoamide); Also their quaternary salts		U.S. Pat. 3,157,595 (1964) German Offen. 1,570,915 (1971)/CA 76, 15356p/	Flocculation of kaolin suspensions
110. Poly(vinylpyridine-HCl), 2-methyl-styrene-maleic acid sodium salt				Flocculation of phosphate mine waters, coal washing water/U.S. Pat. 3,492,225 (1970)/CA 72, 82785h/
111. Styrene-maleic anhydride, and/or itaconic acid anhydride (sodium salts) copolymers		A		Elutriation of digested sewage sludge /U.S. Pat. 3,427,102 (1966)/; Flocculation and separation of precipitated phosphates from wastewaters /U.S. Pat. 3,617,569 (1970)/
112. Polyimideamines	e.g., Quaternary ammonium salt of polyimideamine prepared as reaction product from styrene-maleic anhydride copolymer (mol. ratio 3:1) and 3-dimethyl-aminopropylamine	C	U.S. Pat. 3,507,787 (1970)	Flocculant of aqueous suspensions of finely divided, solid inorganic particles
113. Quaternary ammonium salt of a tertiary amine-containing, imidic reaction product of a monovinyl monomer-maleic anhydride polymer and a polyamine (vinyl-maleic anhydride polymers)	Preferred monomer-styrene	C		Flocculation and oil removal from water surface (water-borne oil slick) /U.S. Pat. 3,415,745 (1968)/

V. FLOCCULANTS BASED ON POLYAMINES AND RELATED SUBSTANCES

A. Polyamines

Chemical Composition	Characteristic Group, and/or Outlined Preparation	Type	Literature References	Examples of Application
114. a) Polyamines	$H_2N(CH_2CH_2NH)_xH$		German Offen. 1,096,836	Flocculation of sewage, removal of

Chemical Composition	Characteristic Group, and/or Outlined Preparation	Type A = Anionic C = Cationic N = Nonionic	Literature References	Examples of Application
			(1955) French Pat. 1,534,146 (1968)	phosphates (lime and polyamines) /U.S. Pat. 3,607,738 (1971)/CA 75, 143817f/; Chemical conditioning of biological sludges for vacuum filtration/J. Walter Pollut. Cont. Fed. 1970, 2(Pt2), R1–R20/CA 72, 124843h/; Neutralization and recovery of baths and solutions in metal industry /German Offen. 2,154,462 (1972) /CA 77, 79317e/; Clarification of sugar juice/U.S. Pat. 3,567,512 (1971)/CA 74, 127946w/
b) Polyalkylenepolyamines	The general formula: $H_2N(C_nH_{2n}NH)_xH$, $x<2$ e.g., Polyalkylenetriamine, triethylenetetramine	C		Clarification of surface waters (with a combination of alkali metal permanganate)/U.S. Pat. 3,483,120 (1960)/; In-line filtration, coated filter aid methods/U.S. Pat. 3,352,424 (1967)/ Coagulation of inorganic particles U.S. Pat. 3,219,578 (1965)/
c) Polyalkylamines		C	Belgian Pat. 644,722 (1964) /CA 63(8), 1965, 9499b/	Polyelectrolyte coatings for filter media (diatomaceous earth filtration of water)
115. Phosphoramide derivatives of polyamines	e.g., Prepared by reacting dialkyl phosphites with polyamines: $(EtO)_2\overset{\overset{O}{\|}}{P}-NH-(CH_2CH_2NH)_2^+H\cdot HCl^-$	C	U.S. Pat. 3,576,741 (1971)	In clarifying of surface waters
116. Quaternized polyamines			Brit. Pat. 1,064,160 (1967) /CA 67, 11997q/	
117. Polyether-amines prepared from: a) 2-chloroethylglycidyl ether-			German Offen. 2,262,284	Dehydration of paper pulps/Switzerland

Chemical Composition	Characteristic Group, and/or Outlined Preparation	Type A = Anionic C = Cationic N = Nonionic	Literature References	Examples of Application
diethylenetriamine and 1,3-bis[(3-chloro-2-hydroxypropyl)dimethylammonium]-2-hydroxypropane dichloride			(1973)/CA 79 (1973), 105738f/ German Offen. 2,127,082 (1973)/CA 76, 101497s/	Pat. 18,818,71 (1971)/
b) Diglycidyl ether, alkanol-amine and alkylenediamine	Prepared by treating of di-glycidyl ether with alkanol-amine (e.g., monoethanol-amine) and alkylenedi-amine (e.g., ethylenediamine)		U.S. Pat. 3,349,053 (1967) /CA 67(26), 1967, 117700c/	As flocculant for aqueous clay sus-pensions
c) Diglycidyl ether-ethylene-diamine hydrochloride co-polymer			U.S. Pat. 3,347,802 (1967) /CA 67(26), 1967, 117694d/	As flocculating agent
118. Polyesteramines				
a) Linear-polyester-backbone quaternary ammonium polyelectrolytes	Prepared by quaternary of reaction products of amines and polyesters	C	U.S. Pat. 3,715,335 (1973) /CA 78 (1973), 148709u/	As flocculant
b) Polyesters containing alkyl-enepolyamine groups			U.S. Pat. 3,470,136 (1969) /CA 71, 113626t (1969)/	As flocculating agent
c) Tertiary aminohydroxyalkyl esters of carboxylic acid polymers			South African 71,05,861 (1972)/CA 78 (1973), 98269c/	As flocculant for industrial water treatment, in papermaking
d) Polycondensation resins, prepared by reaction of di-amines with chloroacetic acid ester and tetramethyl-ene glycol	Polymers of the type: $-[CH_2 \cdot CO \cdot O(CH_2)_4 O \cdot CO\ CH_2 \cdot NH \cdot R \cdot NH]-_n$		German Offen. 2,044,024 (1971)/CA 75, 6646k/	Polymers for use in medicines and for clarifying impure water

B. Condensation Products of Polyamines (and Amines) and Halogen Hydrocarbons

Chemical Composition	Characteristic Group, and/or Outlined Preparation	Type	Literature References	Examples of Application
119. High-molecular-weight poly-amine polyelectrolytes	Prepared, e.g., by conden-sation of polyethylene-amines with dichloro-ethane	C	Spanish Pat. 287,939 (1963)	
120. Condensation products of di-amines with alkyl-(or aralkyl)	Prepared by reaction of compounds:	C	South African Pat. 69,08,949 (1969)/CA 74, 64844p/	Flocculation of municipal and indus-trial wastewaters, coagulant aid (with

Chemical Composition	Characteristic Group, and/or Outlined Preparation	Type A = Anionic C = Cationic N = Nonionic	Literature References	Examples of Application
(α,ω) dichloride	$Me_2NCH_2CH_2NMe_2$, N,N'-dimethylpiperazine $Me_2NCH_2CH_2CHMeNMe_2$ or Me_2NH with equivalent amount of $(ClCH_2CH_2)_2O$		German Offen. 2,048,546 (1971)/CA 75, 37028q/ German Offen. 3,210,308 (1965)	Al salts); Flocculant in papermaking
121. Condensation polymers of alkyl dihalides and polyalkylene polyamines			U.S. Pat. 3,219,578 (1965)	Coagulating dilute inorganic aqueous suspensions
122. Condensation product of a polyhalogenated organophosphate and a polyamine	e.g., By reaction of $(ClCH_2CH_2O)_3PO$ with polyhexamethylenepolyamine		U.S. Pat. 3,542,688 (1970) /CA 74, 43121k/	Flocculant for inorganic and organic aqueous suspensions
123. Copolymer 1,4-butanediol bis(chloroacetate) pentaerythritol tetrakis [β-(dimethylamino)propionate] N,N,N',N'-tetramethylhexamethylenediamine			German Offen. 2,210,433 (1972)/CA 78 (1973), 59244w/	For clarifying aqueous sludge
124. Condensation product of alkyldimethylamines (C_{10-20}) and dichloroethane			USSR Pat. 380,355 (1973) /CA 79 (1973), 83364b/	Sedimentation of high-dispersive mineral particles in water suspensions
125. 2-Methylene-3-butenyl quaternary ammonium monomers and polymers	Prepared, e.g., by polymerization of 2-methylene-3-butenyl ammonium compounds, or by treating poly[2-(chloromethyl)-butadiene] with tertiary ammonium compounds aqueous solutions	C	U.S. Pat. 3,673,164 (1972) /CA 77, 89121a/ U.S. Pat. 3,544,532 (1970)	Flocculation of suspensions in mining operations

C. Various Modified Polymers—Polyamino

Chemical Composition	Characteristic Group and/or Outlined Preparation	Type A = Anionic C = Cationic N = Nonionic	Literature References	Examples of Application
126. Soluble diphenyl ether polymers	A = H or —CH$_2$— Y = Cl, Br, OH or OR R = C$_1$–C$_4$ alkyl reacted with tertiary amine or quaternary ammonium group and Friedel-Crafts catalyst at 0°–85°C		U.S. Pat. 3,316,186 (1967) /CA 67, 11982f/	As flocculants, thickeners and binders
127. Copolymer nonaethyleneglycol bis(chloroacetate)—N,N,N',N'-tetramethylethylenediamine-2,4,6-tris(N,N-dimethylaminomethyl) phenol			German Offen. 2,210,433 (1972)/CA 78 (1973), 59244w/	Clarifying aqueous sludge
128. Acyclic polyamines	Prepared, e.g., by treating of amine with dihaloalkylene		French Pat. 1,449,204 (1966)/CA(16), 1967, 66382v/	Coagulating agents
129. Quaternary ammonium polymer compositions, prepared by epoxidation of polybutadiene followed by reaction of epoxide with tertiary amines			German Offen. 2,141,941 (1972)/CA 77, 35507n/	Flocculating solid particles from their aqueous suspensions
130. a) Guanidine-derived polyelectrolytes	Compounds formed by reacting guanidine-type compounds with polyamines; Guanidine type compounds: e.g., guanidine, dicyandiamide, biguanidine	A	U.S. Pat. 3,617,570 (1971) /CA 76, 114403h/	As water clarifiers, corrosion inhibitors in raw oil processing

Chemical Composition	Characteristic Group and/or Outlined Preparation	Type A = Anionic C = Cationic N = Nonionic	Literature References	Examples of Application
	Polyamines (polyalkylene-polyamines): ethylenediamine, diethylenetriamine, propylenediamine, and/or subst. with alkyl, aryl groups, etc.			
b) Reaction product of cationic polyamine and guanidine derivative			Swiss Pat. 531,362 (1973) /CA 78 (1973), 161521w/	Filter aid
D. Polyamines and Halohydrins				
131. Condensation products of polyalkylene polyamines and epoxyhalides	The condensation polymers are prepared by condensing the "prepolymer" with an α,β-epoxyhalide compound (epichlorohydrin, epibromohydrin); The prepolymers are made from smaller chain polyalkylene polyamines by a condensation reaction with an alkyl dihalide (e.g., ethylene dichloride)	C	U.S. Pat. 3,391,090 (1968)	Flocculant, and/or coagulant—removing of turbidity, color bodies and anionic contaminants from water (often by adding a nucleating agent such as clay)
132. Condensation products of alkylene polyamine and polyfunctional halohydrins (the alkylene polyamine polyfunctional halohydrin polymers)		C		Inline filtration /U.S. Pat. 3,235,492 (1966)/ Removal of organic coloring matter—preconditioning the water to be filtered in a filter aid filtration process
a) Amine-epichlorohydrin polymers	e.g., $PhNH_2$-, $BuNH_2$-epichlorohydrin polymer		Kogyo Yosui 29, 34–8 (1961)	Flocculating agents
b) Polymer of epihalohydrin and polyalkylene polyamine		C	U.S. Pat. 3,523,892 (1970)	Removing of turbidity, color bodies and anionic contaminants from water
c) Polyquaternary resin polymers of alkylene-polyamines (polyquaternary alkylene	Prepared from epihalohydrin and monoalkylamines, e.g., condensation product epi-	C	U.S. Pat. 3,725,312 (1973) /CA 78 (1973), 160654e/ U.S. Pat. 3,741,891 (1973)	As flocculant for sulfuric acid illmeniite slurries, raw sewage

Chemical Composition	Characteristic Group and/or Outlined Preparation	Type A = Anionic C = Cationic N = Nonionic	Literature References	Examples of Application
polyamine resin polymers	chlorohydrin-methylamine is quaternized with dimethyl-sulfate and ethylene oxide (Me sulfate, ethylene-oxide, epichlorohydrin-methyl-amine copolymers)		/CA 79 (1973), 54552f/	
d) Quaternary ammonium derivatives of epichlorohydrin	e.g., Quaternization of epi-chlorohydrin-ethylene oxide (1:1), and/or polyepichloro-hydrin, with Me_3N-type compounds (or pyridine)		German Offen. 2,031,622 (1971)/CA 74, 142628n/	As flocculating agents
e) Amine-modified alkylene oxide polymers	Prepared, e.g., by modifica-tion of polyepichlorohydrin or its copolymers with poly-amines (e.g., ethylenedi-amine), crosslinking agent: ethylene dichloride		German Offen. 2,244,513 (1973)/CA 78 (1973), 160457t/	Dewatering of paper pulp slurries, as flocculant
f) Poly-β-asparaginepolyalkyl-enepolyamine halohydrin resins	Prepared by reacting a poly-alkylenepolyamine with a half ester of maleic acid and partially crosslinking the re-sulting product with a poly-functional halohydrin (e.g., tetraethylenepentamine + monomethyl maleate + epi-chlorohydrin)	C	U.S. Pat. 3,351,520 (1967)	Flocculant for aqueous dispersions or slurries of silica, carbon, clay, bio-logically treated industrial wastes (e.g., textile mill wastes), sewage sludge, pulp slurries, wet strength agents for paper
133. Condensation products of alkyl-ene polyamine and halohydrin			U.S. Pat. 3,408,292 (1968) /CA 70, 14312k/	Inline clarification of water
a) Polymers from epichlorohy-drin and methylamine	Graft polymerizing one or more water-soluble vinyl polymers upon a hydrophilic water-dispersible cationic polyhydroxyalkylene poly-amine	C	U.S. Pat. 3,493,502 (1970) /CA 72, 67751a/ U.S. Pat. 3,567,659 (1971) /CA 74, 142811s/ U.S. Pat. 3,697,370 (1972)	For dewatering and/or settling of aqueous suspensions of finely di-vided solids, sewage sludge, indus-trial process and wastewater, e.g., effluent from iron ore processing, precipitates from neutralization of acidic coal mine waters
b) Quaternary adducts of poly-		C	U.S. Pat. 3,320,317 (1967)	Flocculating agents, particularly ef-

Chemical Composition	Characteristic Group and/or Outlined Preparation	Type A = Anionic C = Cationic N = Nonionic	Literature References	Examples of Application
epichlorohydrins with tertiary lower alkyl amines (polyoxyalkylene polytrialkylammonium chlorides)				fective for improving sedimentation of sewage solids
c) Polyquaternary flocculants-copolymers of sec. amines and epichlorohydrins or di-epoxides	e.g., Epichlorohydrin + mixture of diethylamine and diethylenetriamine		U.S. Pat. 3,738,945 (1973) /CA 79 (1973), 54302z/	Clarification of water, demulsifying agent, filter aid
d) Amino-epoxide polymers and copolymers and their quaternary salts			U.S. Pat. 3,403,114 (1968) /CA 69, 107385f/	As flocculants
e) Polycation flocculation agents based on amino resins	e.g., Reaction products (1:1.01–1.5 molar ratio) of dimethylamine, monoethylamine or monoethylamine and epichlorohydrin, with hexamethylene-pentamine or pentaethylenehexamine	C	Jap. Kokai 72, 42,272 (1972) /CA 78 (1973), 163776g/ Jap. Kokai 72, 49,999 (1972) /CA 78 (1973), 163753x/	Conditioning of sludges, flocculation of dye suspensions, bentonite, porcelaneous clay, and other suspensions
134. Condensation products—prepared by treating of linear 1:1 polyamine-polyacrylamide condensation polymer with epichlorohydrin in ratio 0.2:1			U.S. Pat. 3,305,493 (1967) /CA 66(20), 1967, 86829z/	As flocculant for removal of surface water turbidity, as binder
135. Ammonia-halohydrin condensation polymer			U.S. Pat. 3,408,292 (1968) /CA 70, 14312k/	Inline clarification
a) Ammonia-epichlorohydrin polymer (4:1 molar ratio)		C	U.S. Pat. 3,174,928 (1965)	Coagulation of inorganic solids in low turbidity waters
b) Melamine and epichlorohydrin				

E. Reaction Products of Amines (Poly) and Aldehydes

Chemical Composition	Characteristic Group and/or Outlined Preparation	Type	Literature References	Examples of Application
136. Condensation products of amines and aldehydes:				
a) Aromatic amine aldehyde			Kogyo Yosui 30, 24–32	Flocculation of suspensions

Chemical Composition	Characteristic Group and/or Outlined Preparation	Type A = Anionic C = Cationic N = Nonionic	Literature References	Examples of Application
polymers			(1961)	
b) Amines and formaldehyde	e.g., Formaldehyde and dicyandiamine		German Offen. 1,917,050 (1969)/CA 72, 15367k (1970)/	Filtration of silicates in aqueous suspensions
c) Urea-formaldehyde resin modified with dicyandiamide and diethylene triamine (amino-aldehyde resins)		C		For removal of both inorganic and organic contaminants from wastewater systems, alkyl-benzene sulfonates, phenol derivatives etc. /U.S. Pat. 3,484,837 (1969)/
d) Reaction product of dialkyl-ammonium salt, paraformaldehyde and amide polymer		C	Belgian Pat. 667,873 (1966) /CA 65(3), 1966, 4115g/	Flocculating agents
e) Cationic aniline-formaldehyde resins modified by melamine or by diethanolamine and ethyleneimine (cationic)			Jap. Pat. 71, 10,408 (1971) /CA 76, 49613k/	Separation of organic solids from sewage
f) Urea-formaldehyde condensates, mixture of urea (or guanidine)—aldehyde (e.g., formaldehyde) condensates and polyacrylamide—mixture of inorganic salt with some reaction products of urea resins	e.g., Inorganic salt: NaCl, $Al_2(SO_4)_3$, $FeCl_3$ Resin: urea-formaldehyde, urea-melamine-formaldehyde		Belgian Pat. 549,049 (1956) German Offen. 1,243,646 (1967)/CA 1967(22), 103160q/	Flocculation agents for aqueous suspensions of kaolin, of fixed white suspensions etc.

VI. FLOCCULANTS BASED ON ALKYLENIMINES (AZIRIDINES), ETHYLENE (ALKYLENE) OXIDES

Chemical Composition	Characteristic Group and/or Outlined Preparation	Type A = Anionic C = Cationic N = Nonionic	Literature References	Examples of Application
137. Polyalkylenimines	Prepared by polymerization by alkylenimine monomers (C_1–C_4); e.g., polypropyleneimine	C	U.S. Pat. 3,210,308 (1965) U.S. Pat. 3,203,910 (1965)	Inline clarification /U.S. Pat. 3,352,424 (1967)/ Preventing, controlling and removing deposits of alluvium in cooling water systems/U.S. Pat. 3,288,640 (1966) /CA 66(8), 1967, 29803n/
138. Polyethylenimines	$NH_2(CH_2CH_2NH)_nN$	C	Belgian Pat. 643,454 (1964) U.S. Pat. 3,408,292 (1968) /CA 70, 14312k/ German Offen. 2,064,575	Inline clarification; preventing, controlling and removing deposits of alluvium in cooling water systems /U.S. Pat. 3,288,640 (1966)/;

Chemical Composition	Characteristic Group and/or Outlined Preparation	Type A = Anionic C = Cationic N = Nonionic	Literature References	Examples of Application
			(1972)/CA 77, 118008b/ Jap. Pat. 72, 22,801 (1972) /CA 78 (1973), 75637g/ Jap. Kokai 73, 20,768 (1973) Environ. Sci. Technol. 1973, 7(7), 614–19	Clarification of water from paper mills/Shikizai Kyokaishi 1973, 46(6), 365; /CA 79, (1973), 116447m/Bum. Prom. 1972, (5), 18; /CA 77, 38825a/; Retention agent in papermaking /Tappi 1973, 56(4), 112; /CA 78 (1973), 161107r/; Purification of wastewaters from the polymerization of styrene; Flocculation of wastewaters; Oil-in-water emulsions breaker; Flocculation of starch precipitated Cu from wastewaters; Flocculation and thickening of sludges /Terres Eau 21(55), 1968, 27; /CA 71 (1969), No. 4, 14242n/; Dewatering of inorganic sludge and suspensions together with anionic flocculants; Flocculation of solids and dewatering of sludge with polyalkylene acid-amide dual system/German Offen. 1,920,590 (1969); /CA 72 (1970), 47211n/
139. Polymers from N-substituted ethylenimines and poly-aziridinyl compounds	Prepared by polymerizing in the presence of an acidic catalyst, a compound from the group consisting of ethyleneimine and mono-meric N-subst. ethyleneimines with polydiaziridinyl compounds; substituent, e.g., acyl, ureido, amido, guanidino, carboxyalkyl, carboxy, etc. Polyaziridinyl compounds, e.g.,		U.S. Pat. 3,468,818 (1969) /CA 71 (1969), 102453h/	Coagulation of low turbidity water, flocculation and dewatering of sewage, the addition to paper slurries for the purposes of retaining fine fibers, the breaking of oil-in-water emulsions

Chemical Composition	Characteristic Group and/or Outlined Preparation	Type A = Anionic C = Cationic N = Nonionic	Literature References	Examples of Application
	trisaziridinyl phosphine oxide, dimethyl diaziridinyl silane			
140. Copolymers of ethylenimine and epichlorohydrin			U.S. Pat. 3,294,723 (1966) /CA 66(14), 1967, 55969r/	
141. Carbamoylated polyethylenimine (carbamoyl-polyethylenimine) and its hydrolysis products	Perpared, e.g., by treating of polyethylenimine with NaCNO and HCl; or polyethyleneimine hydrochloride with AgOCN or NaOCN		Jap. Kokai 73,38,878 (1973) /CA 79 (1973), 149268g/	Coagulant of aqueous suspensions at pH 8
			Polymer 1972, 13(4), 187 and 13(11), 552; Kyoto Daigaku Nippon Kagakusen Kenkyusho Koenshu 1972, 29, 45 /CA 79 (1973), 125104d/	Flocculation of aqueous kaolin suspensions
142. Substituted acylated polyimine resins	Prepared, e.g., by partially deacylating of linear poly-(2-ethyl-2-oxazoline) and crosslinking the obtained products, e.g., with epichlorohydrin or diepoxide		U.S. Pat. 3,640,909 (1972) /CA 76, 129141v/	As flocculating agents and retention aid in papermaking
143. Fluorinated copolymers of alkylenimines	Polymeric reaction products of alkylenimines (e.g., ethylenimine, or propylenimine) and fluorinated unsaturated hydrocarbons (e.g., tetrafluoroethylene, hexafluoropropylene, hexafluoro-2-butyne		U.S. Pat. 3,341,476 (1967) /CA 67(20), 1967, 91240a/	As flocculants, e.g., for raw sewage
144. Acryloxyalkyloxyalkylketimines and -aldimines, their polymers, their primary amine monomer and polymers			U.S. Pat. 3,497,485 (1970) /CA 76, 25756h/	Flocculation of aqueous clay suspension, treatment of industrial wastewater, sewage, pulp
145. Copolymers of alkylenimine	Are obtained by an acid cata-		U.S. Pat. 3,579,488 (1971)	Treatment of sewage, flocculation of

Chemical Composition	Characteristic Group and/or Outlined Preparation	Type A = Anionic C = Cationic N = Nonionic	Literature References	Examples of Application
(C_2-C_4) and alkylene sulfide (C_2-C_4)	lyzed copolymerization (catalyst is an organic sulfonic acid), e.g., copolymerization of ethylenimine-ethylene sulfide with toluenesulfonic acid catalyst		/CA 75, 36992n/	aqueous clay suspension
146. Ethyleneimine (aziridine) compounds:				
a) Terpolymers of a C-substituted aziridine, N-substituted aziridine and organic dihalide	Terpolymers are prepared from these monomers in the presence of an inert solvent, e.g., reaction product of ethylene dichloride and ethyleneimine and 1-phenethylaziridine	C	U.S. Pat. 3,498,932 (1970)	Flocculating agents of ore solutions (e.g., magnetite, hematite ore)
b) Homopolymers of N-(2-hydroxyethyl) aziridines and N-(2-thioethyl) aziridines			U.S. Pat. 3,752,854 (1973) /CA 79 (1973), 126939n/	Flocculating agents
147. Polyethylene oxide		N		Flocculant in the slime flotation process, e.g., talc, limonite etc. /Suiyokai - Shi 1968, 16(6), 289; /CA 69, 21235v/; Flocculation of coal slurries/Izv. Dnepropetrovk. Gorn. Inst. 1971, 57, 124/, /Ugol. Ukr. 11(3), 43 (1967)/; Floccuation–centrifugation treatment of flotation tailings/Tr. Inst. Obogashch. Tverd. Gornych Iskop. 1971, 1(1), 111; /CA 79 (1973), 33453j/
148. Oxyalkylated surfactants	Most effective are those which contain more than one alkylene oxide, e.g., alkyl phenol-ethylene oxide condensation products.	N	U.S. Pat. 3,194,758 (1965)	Agglomerating finely divided solids aqueous medium, clarifying effluents from deinking processes /U.S. Pat. 3,354,028 (1967)/

Chemical Composition	Characteristic Group and/or Outlined Preparation	Type A = Anionic C = Cationic N = Nonionic	Literature References	Examples of Application
	In general, they may be obtained by condensing a polyglycol ether containing the required number of alkyleneoxy groups or an alkylene oxide (such as propylene oxide, butylene oxide, or preferably ethylene oxide) with a suitable alkylphenol; The general formula: $R(CHR_1CHR_1O)_nH$ R = the residue of a suitable alkyl phenol R_1 = H or lower alkyl n = 2-100 (4-30) Suitable nonionic detergent is an ethylene oxide adduct of dodecylphenol (n = 8-15), or a condensation product of nonylphenol and ethylene oxide of the structural formula: $C_9H_{19}-\phi-O-(CH_2CH_2O)_n-H$ n = 8-15.			

VII. FLOCCULANTS BASED ON POLYAMIDES

Chemical Composition	Characteristic Group and/or Outlined Preparation	Type	Literature References	Examples of Application
149. Polyamides (soluble in acid)	Prepared, e.g., by reaction of polymeric aliphatic acid and polyalkylene polyamine in excess. Aliphatic acids: polymerization of vegetable oils—tall oil, soap stock. Polyalkylene-polyamines: diethylenetriamine, triethylenetetramine, tetraethylenepentamine, di-1,3-propantriamine, tri-1,3-propantetra-	C	South African 68, 05,823 (1969)/CA 71, 114420h (1969)/ Brit. Pat. 906,854	Flocculation of industrial wastes, dewatering of suspensions

Chemical Composition	Characteristic Group and/or Outlined Preparation	Type A = Anionic C = Cationic N = Nonionic	Literature References	Examples of Application
	amine, di-1,2-propantriamine			
150. Polymeric amide—dialkyl ammonium salt—paraformaldehyde flocculant	Prepared, e.g., from polyacrylamide (1% amide groups hydrolyzed), dimethylammonium oxalate, paraformaldehyde and sodium carbonate	C	U.S. Pat. 3,367,918 (1968)	Flocculation of aqueous inorganic suspensions (coal fines, silica, clay) and organic matter (polysaccharides, cellulose fiber)
151. Polyamide-epichlorohydrin			French Pat. 1,570,656 (1969) /CA 72 (1970), 45618k/	Flocculation agent for titanium sulfate liquors
152. Polymeric N-carbamoylsulfoamides	e.g., β-(methacryloyloxy)-ethyl N-(tolylsulfonyl)-carbamate-methyl methacrylate copolymer; Maleinidomethyl N-(β-chloroethylsulfonyl)carbamate-styrene copolymer; Acrylonitrile-N-acryloyl-N-(p-tolylsulfonyl) urea copolymer		German Offen. 2,027,466 (1971)/CA 76, 114137z/	As binder components for light-sensitive layers, retention aid, dispersants of pigments and flocculants
153. Amino-substituted polyamides		C		Clarification of surface water (river water)/U.S. Pat. 3,483,120 (1966)/
154. Water-soluble polymers of poly-aminotriazole and succinic acid hydrazide			Kogyo Yosui 28, 35–8 (1961)	Flocculation of aqueous active carbon, kaolinite, Fe_2O_3 etc.
155. Reaction product of sulfuric acid and allophane			U.S. Pat. 3,535,259 (1970) /CA 74, 45985n/	As flocculant

VIII. FLOCCULANTS BASED ON POLYSTYRENE

Chemical Composition	Characteristic Group and/or Outlined Preparation	Type	Literature References	Examples of Application
156. Polyelectrolytes, prepared by chloromethylation following amination or sulfonation of styrene copolymers	e.g., Styrene-diamine copolymers		USSR Pat. 303,326 (1971) /CA 75, 152367m/	Chemical conditioning of biological sludges for vacuum filtration/J. Water. Pollut. Contr. Fed. 1970, 2(Pt 2), R1-R20; /CA 72, 124843h/
a) Sulfonated polystyrene		A	U.S. Pat. 3,336,271 (1967)	Treatment of raw influent sewage

Chemical Composition	Characteristic Group and/or Outlined Preparation	Type A = Anionic C = Cationic N = Nonionic	Literature References	Examples of Application
and/or Na-salt)				/U.S. Pat. 3,479,283 (1969)/
b) Aminated chloromethylated polystyrene		C	Zh. Prikl. Khim. 39, 2679 (1966) Zh. Prikl. Khim. 34, 2430 (1961) Czech. Pat. 100,709 (1961) Sb. Nauch. Tr. Kuzbas. Politekh. Inst. 1971, No. 34, 84–8 /CA 78 (1973), 62023d/	Clarification of water
157. Polyvinylbenzyl compounds				
a) Polyvinylbenzyltrimethylammonium chloride		C	Environ. Sci. Technol. 1973, 7(7), 614–19	Purification of oxytetracycline solutions/Khim. Farm. Zh. 1(8), 52 (1967); /CA 68, 43131q/ Dewatering of suspensions of organic solids (with anionic polyelectrolytes)/U.S. Pat. 3,259,570 (1966)/ Flocculation of precipitated Cu-starch xanthate from wastewater, clarification of surface waters/U.S. Pat. 3,483,120 (1969)/ Inline clarification, precoat filtration (coating of filter aid media)/U.S. Pat. 3,352,424 (1967)/
b) Polyvinylbenzyltertiary amines		C		Inline clarification, coating of filter aid media/U.S. Pat. 3,352,424 (1967)/
c) Vinylbenzylsulfonium polymers		C		Inline clarification, coating of filter aid media/U.S. Pat. 3,352,424 (1967)/
d) Poly(vinylbenzylsulfonium) halides		C	U.S. Pat. 3,078,259 (1963) U.S. Pat. 3,130,117 (1964) U.S. Pat. 3,216,979 (1965) M.S. Matsch, et al: J. Appl. Pol. Sci. 13, 721 (1969)	
158. Copolymers of the type:	$-[CH_2CH(CONH_2)]_{\overline{x}}$ $-[CH_2CH(C_6H_4CH_2A)]_{\overline{y}}$			

Chemical Composition	Characteristic Group and/or Outlined Preparation	Type A = Anionic C = Cationic N = Nonionic	Literature References	Examples of Application
	Where A = $-^+NR_3$ $-^+PR_3$ $-^+SR_2$		U.S. Pat. 3,068,213 (1962) U.S. Pat. 3,068,214 (1962) U.S. Pat. 3,060,156 (1962)	
159. Vinyl aromatic sulfonate polymers (homo- and copolymers)	Polyvinyl aromatic sulfonates can be prepared by the sulfonation of an essentially linear, high molecular weight vinyl aromatic polymer (such as polystyrene, polyvinyl toluene, poly-α-methylstyrene and mixtures thereof, particularly sodium polystyrene sulfonate copolymers can be prepared by the vinyl polymerization of the major proportion of a vinyl aromatic sulfonate monomer with a minor proportion of a monomer copolymerizable therewith such as acrylamide, acrylonitrile, styrene, etc.	A	U.S. Pat. 3,340,238 (1967) /CA 70(20), 1967, 91248j/ U.S. Pat. 3,336,271 (1967) /CA 67(10), 1967, 74156k/	Thickening and dewatering of organic waste sludge (sedimentation and filtration, particularly raw and digested municipal sewage/U.S. Pat. 3,300,407 (1967)/; Elutriation of digested sewage sludge /U.S. Pat. 3,247,102 (1966)/; Clarification of red mud slurries obtained by alkali leaching of aluminiferous ores/U.S. Pat. 3,194,757 (1965)/; Dewatering of suspensions of organic solids (with cationic polyelectrolytes) /U.S. Pat. 3,259,570 (1966)/
160. Condensed polycyclic compounds, e.g., cracked asphalt, containing 4–5 of condensed aromatic cores subjected to cationic methylation			Jap. Pat. 72-26,580 (1972) /CA 78 (1973), 33668n/	Flocculants for wastewater, sewage

IX. FLOCCULANTS BASED ON HETEROCYCLIC POLYMERS

Chemical Composition	Characteristic Group and/or Outlined Preparation	Type	Literature References	Examples of Application
161. a) Polyvinyl pyridine hydrochloride		C		Flocculation of suspensions having negative charge; flocculation of kaolinite aqueous suspensions/Kogyo Kagaku Zashi 69(6), 1192 (1969); /CA 66, 116327d/
b) Poly(1,2-dimethyl-5-vinyl-pyridinium methyl sulfate)			J. Polym. Sci. Part A-2, 1969, 7(1), 27–36	Flocculation of crystalline silica

	Chemical Composition	Characteristic Group and/or Outlined Preparation	Type A = Anionic C = Cationic N = Nonionic	Literature References	Examples of Application
162.	Polyelectrolytes derived from N-methylvinylpyridinium sulfate	Copolymers with styrene, vinyltoluene, methacrylic acid	C	Kagaku Tokogyo (Osaka) 40(1), 17–34 (1966)/CA 65(2), 1966, 17741f/	As flocculants
163.	Polyacrylamide + polyvinylpyridine·HCl	Dual systems		Jap. Pat. 68-18,731 (1969) /CA 72, 47908d (1970)/	Coagulation of suspended microparticles (having negative charge)-
164.	Vinyl pyridine and epichlorohydrin polymers			USSR Pat. 201,982 (1966) /CA 69, 61826v/	As flocculants
165.	Ammonium group-containing vinyl polymers	e.g., Poly(1,2-dimethyl-5-vinylpyridinium methylsulfate), poly(β-methacryloyloxyethyltrimethylammonium methyl sulfate)			Flocculants for activated sludge German Offen. 2,046,920 (1971); /CA 75, 25020t/
166.	Cyclic amidine polymers (heterocyclic polymers containing a cyclic amidine group in the main polymeric chain)	Are formed by reacting cyanohydrins with cyclic amidine-forming polyamines to form polyamino nitriles and reacting this nitrile intermolecularly with itself or with a polyamine under cyclic amidine-forming conditions Or formed by the intermolecular reaction of a cyanoethylated cyclic amidine-forming polyalkylene polyamine	C	U.S. Pat. 3,576,740 (1971) /CA 75, 22035r/ U.S. Pat. 3,450,646 (1969) /CA 71, 61894g/ U.S. Pat. 3,509,046 (1970) /CA 73, 46357g/	Flocculants for water suspensions, water clarifiers, oil-in-water emulsion breakers, corrosion inhibitors
167.	Polymers of vinylimidazoline	Obtained by reaction of excess ethylene-diamine with a polyacrylonitrile e.g., bisulfate of a 2-vinyl-imidazoline polymer, poly-(2-vinyl-2-imidazoline), poly(2-vinyl-4(or 5)-methyl-	C		Removal of anionic water-soluble materials (anionic detergents, decay products, humic acid)/U.S. Pat. 3,300,406 (1967); /CA 67(2), 1967, 5621w/; Treatment of aqueous suspensions of finely divided water-insoluble solid

Chemical Composition	Characteristic Group and/or Outlined Preparation	Type A = Anionic C = Cationic N = Nonionic	Literature References	Examples of Application
	2-imidazoline]			materials (minerals or organic matter), filtration and clarification of biological sewage sludges/U.S. Pat. 3,288,707 (1966)/
168. Polymer 2-vinyl-3,4,5,6-tetra-hydropyrimidine hydrochloride	Obtained by reaction of excess trimethylenediamine with a polyacrylonitrile (molecular weight 600,000) so that the nitrile groups are converted to tetrahydropyrimidine groups	C	Neth. Appl. 6,515,018 (1966)/CA 66 (16), 1967, 66379z/	Purification of municipal and industrial wastewater, removal of anionic water contaminants (in combination with bentonite); Flocculation of clay and microorganisms
169. Polymers containing 5 or 6 membered saturated heterocyclic rings				
a) Polyvinylpyrrolidone				Coagulation of finely divided solids (predominantly inorganic); Coagulation of phosphate mine water, coal washery waters/U.S. Pat. 3,492,224 (1970); 3,492,225 (1970); CA 72, 82784g/
b) Poly(2-piperidylidenemethylene)	Prepared by hydrogenation and acidification of cyano-ethylene-polyacrylonitrile		Cercet. Miniere 1968 (Pub. 1970), 11, 393–5/CA 74, 64809f/	Flocculant for nonmetallic substances and coal sludges
170. Polyvinyl pyrrolidone (cross-linked polyelectrolytes)	a) Crosslinked, e.g., with N,N-dialkyl acrylamide b) In conjunction with poly-anionic materials (synergistic action) e.g., glue, gelation, cellulose derivatives (e.g., carboxymethyl-cellulose), natural gums (e.g., guar gum)	C	U.S.. Pat. 3,235,490 (1966) /CA 64(9), 1966, 12978h/	Flocculation of slimes
171. Polyglycidyl polymers (homo- and copolymers, derivatives-glycidyl ethers)	e.g., N-(2,3-epoxy-1-propyl)-morpholine	N	U.S. Pat. 3,519,559 (1970) /CA 73, 69710f/	Water clarifiers, oil-in-water demulsifier

Chemical Composition	Characteristic Group and/or Outlined Preparation	Type A = Anionic C = Cationic N = Nonionic	Literature References	Examples of Application
X. FLOCCULANTS BASED ON POLYMERS OF DIALLYLAMMONIUM (AND DIALLYLDIALKYLAMMONIUM) SALTS				
172. Modified polyamines (polyquaternary compounds), prepared by polymerizing a quaternary ammonium monomer or by alkylation of already formed polyamine compounds (polymers of polydialkyldiallyl ammonium salt)	e.g., Diallyldimethylammonium chloride		Neth. Appl. 6,514,783 (1966)/CA 68, 3330x/ U.S. Pat. 3,288,770 (1966) U.S. Pat. 3,461,163 (1969) U.S. Pat. 3,472,740 (1969) Fr. Demand. 1,494,438 (1967)	Clarification of water containing residual chloride/U.S. Pat. 3,409,547 (1968)/; Filtration of liquids
173. Addition compounds (and their quaternary salts) of diallylamine with acrylic derivatives	e.g., Diallylmethyl (2-carbamoylacryloyl) ammonium chloride; Acrylic derivatives: acrylates, acrylonitrile, acrylamide Quaternary salts: diallylmethyl(cyanoethyl)ammonium Me sulfate, diallylmethyl(2-methoxycarbamoylethyl) ammonium Me sulfate, diallyl (β-hydroxyethyl) (2-carbamoylethyl) ammonium chloride	C	Neth. Appl. 6,607,032 (1966)/CA 66(24, 1967,	Flocculating agents
174. Diallylamines (and salts) e.g., diallylbenzylamine, diallylmethylamine, diallylethylamine		C		Flocculant for wastewaters, (digested, activated) sludge filtration/U.S. Pat. 3,171,805 (1965)/
175. Polymers comprising groups of repeating units derived from diallylamine and quaternary ammonium monomers containing groups condensed through Michael addition reaction from a vinyl type activated double bond compound	e.g., Homopolymers of dialkyl methyl(β-propionamido)ammonium chloride	C		Flocculation of wastewaters/U.S. Pat. 3,412,019 (1968)/

Chemical Composition	Characteristic Group and/or Outlined Preparation	Type A = Anionic C = Cationic N = Nonionic	Literature References	Examples of Application
		XI. FLOCCULANTS BASED ON POLYSULFONES AND POLYSULFINES		
176. High molecular weight polysulfones (condensation products of sulfur dioxide and ethylenically unsaturated monomers in the presence of suitable catalyst)	e.g., Ethylenic monomers: mono- and diolefins such as ethylene, propylene, cyclohexene, styrene, butadiene, isoprene; Polyfunctional monomers such as alkyl alcohol, alkyl ether, acrylic acid, acrylonitrile copolymerization of a limited amount of a water-soluble sulfoester of an α-methylenecarboxylic acid with a monoallyl ether of a polyoxyethylene glycol (hereafter referred to as alkylglycolether) and sulfur dioxide; These polysulfones are characterized by receiving ethylenesulfone moieties; The flocculant activity is related to the sulfoester content	A	U.S. Pat. 3,308,102 (1967) /CA 66(20), 1967, 86118s/	As flocculating agent
177. Polymers of unsaturated sulfines	Include: both α-ethylenically unsaturated sulfine homopolymers as well as copolymers thereof with an α-ethylenically unsaturated thio-ether or with an α-ethylenically unsaturated comonomers; The polymers of an α-ethylenically unsaturated sulfine with both an α-ethylenically unsaturated thio-ether and with an α-ethylenically unsaturated comonomer (terpolymers, e.g., with acrylamide, vinylbenzene,	C	U.S. Pat. 3,214,370 (1965)	Flocculation of negatively charges colloids, aqueous dispersion or slurries of silica, carbon, clay; Biologically treated industrial wastes (textile mill wastes, etc.), sewage sludge, white water, etc.

Chemical Composition	Characteristic Group and/or Outlined Preparation	Type A = Anionic C = Cationic N = Nonionic	Literature References	Examples of Application
	N-vinylpyrrolidone, N-alkyl-N-vinylamide)			

XII. MIXTURE OF FLOCCULANTS (SYNERGISTIC COMBINATION)

Chemical Composition	Characteristic Group and/or Outlined Preparation	Type	Literature References	Examples of Application
178. Enhancing effects of cationic polymers	Very rapid settling and high clarity in a minimum of time are obtained by first mixing into the suspension a flocculant such as CMC, guar gum or gelatin and then adding a cationic flocculant such as dimethyl dioleyl quaternary ammonium chloride or diallyl dihydrogenated tallow quaternary ammonium chloride		U.S. Pat. 3,165,465 (1965)	Coagulation and settling of aqueous suspensions of finely divided minerals
179. Mixture of anionic flocculant and amino-polymer (3:1–50:1, preferably from 5:1–15:1)	Anionic polymer flocculants, e.g., homo- and copolymers of alkali metal styrene sulfonates, acrylates, methacrylates, alkali metal and ammonium salts of high copolymers of styrene and subst. styrenes with maleic acid amino polymers, e.g.,: homo- and copolymers of N-vinyl pyridine, aminoalkylacrylates, polyalkylene polyamines, polyethyleneimines			Flocculation of wastewaters/U.S. Pat. 3,539,520 (1970)/
180. Phyllosilisates-dispersed and peptized in aqueous medium together with a chemical compound	e.g., A: kaolinite, saponite B: polyacrylamide, polyacrylonitrile, tannins, glue		South African 68, 00841 (1968)/CA 70 (1969), 50401x/ U.S. Pat. 3,511,778 (1970)	Improvements in the treatment of aqueous materials

Chemical Composition	Characteristic Group and/or Outlined Preparation	Type A = Anionic C = Cationic N = Nonionic	Literature References	Examples of Application
181. Pyrophyllite and inorganic salts			U.S. Pat. 3,350,303 (1967) /CA 43086d/	For clarifying water
182. Composition of clay (A) and synthetic organic polymer (B)	e.g., A: bentonite, saponite B: poly(vinylmethylethyl-maleic anhydride)		U.S. Pat. 3,130,167 (1964)	Coagulation of phosphate mine waters, coal dust
183. Coagulant aid compositions based on the matrix associated with phosphatic materials or, correspondingly on phosphate clays derived from low grade phosphatic materials (rich in phosphorus pentoxide) in conjunction with known coagulant aids (such as polyelectrolytes and inorganic salts)	e.g., Compositions of ferric chloride (35), matric (59) and a polyelectrolyte vegetable gum (6); Polyelectrolytes: suitable vegetable gums (agar-agar, alginates, guar type gums); Synthetic products: poly-acrylamides		U.S. Pat. 3,515,666 (1970) /CA 73, 38411e/	Coagulant in the clarification of water (low pH values)
184. Starch phosphate + flocculant based on polyacrylamide (e.g., Polyteric AS 3)			Brit. Pat. 1,314,431 (1973) /CA 79 (1973), 80657p/	Flocculation of tailings from coal washing
185. Starch and NaOH (20%)			Jap. Pat. 72-20,601 (1972) /CA 77 (1972), 166786c/	Aqueous slurries of $Fe(OH)_2$ and $Mg(OH)_2$
186. Mixtures of vinyl polymer (e.g., polyvinylacetate), resin (e.g., colophony) and water, or organic solvent	As flocculant used with inorganic salts (e.g., alum)		Jap. Pat. 72,21,402/CA 78 (1973), 128331m/	
187. Mixture of tetraethylenepentamine and Na-montmorillonite			French Pat. 1,539,569 (1968)/CA 71, 84385f/	Removal of oil droplets from water
188. Polyacrylamide-aminoplast resin mixtures (synergistic effect)	In a weight ratio of from 1:9 to 9:1	C		Useful for processing minerals in aqueous systems/U.S. Pat. 3,509,021 (1970)/;

Chemical Composition	Characteristic Group and/or Outlined Preparation	Type A = Anionic C = Cationic N = Nonionic	Literature References	Examples of Application
189. Sodium polyacrylate and NaOH				As pigment retention aids in paper-making In paper mills for cellulose settling in "white water"/Kogyo Josui 1968, No. 122, p. 351, CA 73, 132140n/
190. a) Synthetic organic latex (cationic, anionic, nonionic) b) Synergistic combination of alum plus a synthetic cationic organic latex	These latices are water-based emulsions formed by the free radical-induced emulsion polymerization of olefins or dienes in the presence of water and cationic or anionic or nonionic organic emulsifying agent	N,C,A	U.S. Pat. 3,485,752 (1969) U.S. Pat. 3,493,501 (1970) /CA 72, 70451h/	Nonionic: treatment of aqueous inorganic industrial process and wastewater suspensions (coal washing, ore processing), aqueous suspensions of chlorinated polyethylene and polyvinyl chloride; Cationic: flocculant in the clarification of water/U.S. Pat. 3,493,501 (1970)/ Phosphate removal from water by precipitation and flocculation/U.S. Pat. 3,453,207 (1969)/

Key to Abbreviations of Journals

ATIP
    Association Technique de l'Industrie Papetiere Bull. (France)

Angew. Makromol. Chem.
    Angewandte Makromolekulare Chemie (Basel, Switzerland)

Ambio
    Ambio (Stockholm, Sweden)

Aust. Chem. Eng.
    Australian Chemical Engineering

Appl. Microbiol.
    Applied Microbiology (Baltimore, Md., USA)

Bum. Prom.
    Bumazhnaya Promyshlennost (Moscow, USSR)

Cercet. Miniere
    Studii si Cercetari de Miniere (Bucharest, Rumania)

Cvet. Metal.
    Tsvetnyie Metally (Moscow, USSR)

Chem. Tech.
    Chemische Technik (Leipzig, East Germany)

Chim. Prom. (Khim. Prom.)
    Khimicheskaya Promyshlennost (Moscow, USSR)

Environ. Health (India)
    Environmental Health (India)

Environ. Sci. Technol.
    Environmental Science & Technology (Easton, Pa., USA)

Erzmetall
    Erzmetall (Stuttgart, West Germany)

Energetik
    Energetik (Moscow, USSR)

God. Vissh. Minno-Geol. Inst., Sofia
    Godishnikh na Vissh. Minno-Geologicheskiya Institut (Sofia, Bulgaria)

Gas, Woda Sanit.
    Gaz, Woda i Technika Sanitarna (Warsaw, Poland)

Ind. Wastes
    Industrial Wastes (Scranton Publishing Co., Inc., (Chicago, III., USA)

Izv. Vysh. Ucheb. Zaved. Khim. Khim, Technol.
    Izvestiya Vyshchikh Uchebnykh Zavedenij (Ivanovo), Khimiya i Khimicheskaya Technologiya (USSR)

Izv. Dnepropetrovsk. Gorn. Inst.
    Izvestiya Dnepropetrovskogo Gornogo Instituta (USSR)

Jour. Soc. Chem. Ind.
>Journal of the Society of Chemical Industry (London, G.B.)

J. Appl. Polym. Sci.
>Journal of Applied Polymer Science (Interscience Publ., New York, N.Y., USA)

JAWWA
>Journal of the American Water Works Association (Baltimore, Md. - New York, N.Y., USA)

J. Water Pollut. Contr. Fed.
>Journal Water Pollution Control Federation (Washington, D.C., USA)

J. Polym. Sci.
>Journal of Polymer Science (New York - Amsterdam)

Khim. Farm. Zh.
>Khimiko-farmatsevticheskij Zhurnal (Moscow, USSR)

Kozh.-Obuv. Prom.
>Kozhevenno-obuvnaya promyshlennost SSSR (Moscow, USSR)

Kharchova Prom.
>Kharchova Promyslovist (Kiev, USSR)

Kolyma
>Kolyma (USSR)

Kem. Ind.
>Khemiya u Industriyi (Zagreb, Yugoslavia)

Koks Khim.
>Koks i Khimiya, Kharkov (Moscow, USSR)

Keram. Z.
>Keramische Zeitschrift (Lübeck, West Germany)

Liteinoe Proizvod.
>Liteinoe Proizvodstvo (Moscow, USSR)

Nauch. Tr., Karagand. Territ. Otd. Vost. Nauch.-Issled. Inst.
>Nauchnye Trudy Nauchno-Issledovatelskogo Instituta Karagandskoi Territorii, Otdelenia Vostok (USSR)

Nauch. Tr., Tashken. Univ. (Nauch. Tr., Tashkent. Gos. Univ)
>Nauchnye Trudy, Tashkent. Gosudarstvennogo Universiteta (Tashkent, USSR)

Obogashch. Polez. Iskop.
>Obogashcheniya Poleznykh Iskaopaemykh (USSR)

Ochistka Stochnykh Prir. Vod
>Ochistka Stochnykh Priorodnykh Vod (Edited by Bruok-Levinson, T.C., Nauka i Tekhnika, Minsk, Belorus. SSR)

Ochistka Povtornoe Izpolz. Stochnykh Vod Urale
>Ochistka i Povtornoe Ispolzovannie Stochnykh Vod Urale (SSSR)

Obogash. Rud.
>Obogashchenie Rud. Nauchnotekhnicheskij Informatsijnyj Bulletin (Leningrad, USSR)

Przem. Chem.
> Przemysl Chemiczny. (Warsaw, Poland)

Probl. Ispolz. Okhr. Vod. Resur.
> Problemy Ispolzovaniya i Okhrany Vodnykh Resurov (Ed. Gatillo, P.D., Nauka i Tekhnika, Minsk, Belorus. SSR)

Paliva
> Paliva (Prague, Czechoslovakia)

Pure Appl. Chem.
> Pure and Applied Chemistry (London, G.B.)

Pulp Paper Mag. Can.
> Pulp and Paper Magazine of Canada (Montreal, Canada)

Proc. Queens. Soc. Sugar Cane Technol.
> Proceedings of Queensland Society of Sugar Cane Technology (Australia)

Rev. Chim. (Bucharest)
> Revista de Chimie (Bucharest, Romania)

Rudy Metale Niezhelaz.
> Rudy i Metale Niezhelazne (Poland)

Rev. Minelor
> Revista Minelor (Bucharest, Romania)

Sklarm Keram.
> Sklar a Keramik (Czechoslovakia)

Sb. Nauch. Tr., Nauch.-Issled. Proekt. Inst. Obogashch. Aglomer. Rud., Chern. Met.
> Sbornik Nauchnykh Trudov, Naucho-Issledovatelskogo i Proektnogo Instituta Obogashcheniya Aglomer. Rud, Chernaya Metalurgiya (USSR)

Sanit. Tech.
> Sanitarnaya Tekhnika (USSR)

Sin. Vysokomol. Soedin.
> Sintez Vysokomolekulyarnykh Soedinenij (Tashkent, USSR)

Sb. Tr., Gos. Nauk.-Issled. Inst.Svet.Metal.
> Sbornik Trudov, Gosudarstvennogo Naucho-Issledovatelskogo Instituta Tsvetnykh Metallov (USSR)

Sb. Tr. Ukr. Nauch.-Issled. Inst. Tsellyl. Bum. Prom.
> Sbornik Trudov Ukrain. Nauchno-Issledovatelskogo Instituta Tsellyulozno-Bumazhnoi Promyshlennosti (USSR)

Soc. Chem. Ind. (London) Monogr.
> Society of Chemical Industry (London) Monogr.

Tr. Inst. Obogashch. Tverd. Gornych Iskop.
> Trudy Instituta Obogashcheniya Tverdykh Goryuchikh Iskopaemykh (Lyubertsy, USSR)

Tappi
> Tappi (Technical Association of the Pulp and Paper Industry, Easton, Pa. USA)

Terres Eaux
    Terres Eaux (France)

Tr. Bakinsk. Fil. Vses. Nauch.-Issled. Inst. Vodoshabzh. Kanaliz., Gidrotekh. Sooruzh., Inzh. Gidrogeol.
    Trudy Bakinskogo Filiala Vsesoyuznogo Nauchno-Issledovatelskogo Insitituta. Vodosnabzhenie, Kanalizatsiya, Gidrotekh. Sooruzheniya i Inzh. Gidrogeol. (USSR)

Tr. Inst. Gornych. Iskop., Moscow
    Trudy Instituta Goryuchikh Iskopaemykh (Moscow, USSR)

Tr. Nauchnoissled. Inst. Vodosnab. Kanaliz, Sanit. Tekh.
    Trudy Nauchnoissled. Inst. Vodoshabyvanye, Kaniliz. Sanit. Tekh. (Bulgaria)

Ugol (Ukr.)
    Ugol Ukrainy (Kiev, USSR)

Vysokomol. Soedin.
    Vysokomolekulyarnye Soedineniya (Moscow, USSR)

Vod. Hosp.
    Vodni Hospodarstvi (Prague, Czechoslovakia)

Vesti Akad. Nauk Belorus. SSR, Ser. Chim. Nauk
    Vestnik Akademii Nauk Belorus. SSR, Ser. Khim. Nauk (Belorussian SSR)

Vodosnabzh.-Sanit. Tech. (Tekh)
    Vodosnabzhenie i Sanitarnaya Tekhnika (Moscow, USSR)

Vuglishta
    Vuglishta (Bulgaria)

Vodosnabzh. Kanaliz. Gidrotekh. Sooruzheniya
    Vodosnabzhenie, Kanalizatsia, Gidrotekh. Sooruzheniya (USSR)

Vop. Technol. Orab. Vody, Prom. Pitevogo Vodosnabzh.
    Voprosy Tekhnologii Obrabotky Vody, Promyshlennost Pitevogo Vodosnabzheniya (USSR)

Vzaimodejstvie Vodorastvorim. Polielektrolt. Dispersn. Sist.
    Vzaimodeistvie Vodorastvorimmykh Polielektrolitov s Dispersnymi Sistemami (Akademie Nauk Uzbekskoi SSR, Institut Khimii, Izdatelstvo "FAN", Taskent)

Wasser, Luft, Betr.
    Wasser, Luft and Betrieb (Wiesbaden, West Germany)

Z. Prikl. Khim (Leningrad)
    Zhurnal Prikladnoi Khimii (Leningrad, USSR)

Z. Prikl. Khim.
    Zhurnal Prikladnoi Khimii (Moscow, USSR)

Polymer (Japan)

Staerke (Stuttgart, West Germany)

Table   2

Commercial Flocculants
and Their Characteristics
(Commercial Polyelectrolytes)

Flocculant Trade Name	--- Chemical Composition ---				Bulk Density, g/l	Molecular Weight, x 10^6	---- Water Solution Properties ----				Optimum Range of Action, pH	Maximum Work Solution Conc. Recommended, %	Maximum Conc. Recommended for Potable Water Use (by US PHS), mg/l	Manufacturer and/or Distributor (Table 4)
							Solution, %	Viscosity, cp	Density, g/ml	pH				
1	2	3	4	5	6	7	8	9	10	11	12	13	14	15
ACCOFLOC 212			Syn	PAM										65
305			Syn	PAM										65
ACRYSOL A-1	A		Syn	PA										
A-3	A		Syn	PA										
A-5	A		Syn	PA										
ACTI-FLOC														44
AEROFLOC 548	A		Syn	PN	432–592		1	30	1	8.7	WR			5, 25
550	A	P (L 10%)	Syn	PN					1.12	12		20		5, 25
552			Syn	PN										5, 25
570		P	Syn	PAM										5, 25
3000		P	Syn	PAM										5, 25
S-3019	C	P	Syn	XP										5, 25
S-3100														5, 25
3171		P	Syn	PAM										5, 25
3425	A	P	Syn	PN	384–496		1	4–6	1	10–11				5, 25
3453	A	P	Syn	PAM	528–624	5	1	125–250	1	5		1		5, 25
AEROSOL C-61	C	Pa	Syn	AG				Viscous	1	10.4		$<$1		5
AFCOLAC S 100														44
AGRILON N														6
AKRYFLOK K10	C		Syn											26
K50	C		Syn											26
K100	C		Syn											26
AKRYNAX	A	P	Syn	PA										26

Flocculant Trade Name	Chemical Composition				Bulk Density, g/l	Molecular Weight, $\times 10^6$	Water Solution Properties				Optimum Range of Action, pH	Maximum Work Solution Conc. Recommended, %	Maximum Conc. Recommended for Potable Water Use (by US PHS), mg/l	Manufacturer and/or Distributor (Table 4)
							Solution, %	Viscosity, cp	Density, g/ml	pH				
1	2	3	4	5	6	7	8	9	10	11	12	13	14	15
ALFLOC 370													0.5	62
980	C		Syn											62
990	A		Syn											62
ALL-FLOK-ADE													3.0	4
No. 6													1.0	4
ALLSTATE No. 2													3	2a, 4
No. 6													1	2a, 4
ALGIN	A	P	Nat	AL										76
ALGINAT 410			Nat	AL										76
ALCHEM COAGU-AID 252													5	31, 31a
265													1	31, 31a
AMERFLOC 265													1	31b
275													1	31b
307													1	31b
AMF			Syn	PAM-CO										98
AQUAFLOC 403	A	L		XP			Orig	100	1.12	9.6		4:1 (D)		27
405	C	L	Syn	XP			Orig	1,000	1.06	6.8		10:1 (D)		27
407	N	L		XP			Orig	10,000	1.0	4.3		10:1 (D)		27
408	A	L	Syn	XP			Orig	1,500	1.01	3–4		10:1 (D) 1:5 (D)	50	26
409	A	P	Syn	PS	448		1	2,350	1	4–5		<1%	1	27
410	C	L	Syn	XP			Orig	1,000	1.03	10.7	WR	1:10 (D) 1:5 (D)		27
411	A	P	Syn	PS	720		2	480	1	4–5		2	2	27

Flocculant Trade Name	--- Chemical Composition ---				Bulk Density, g/l	Molecular Weight, $\times 10^6$	---- Water Solution Properties ----				Optimum Range of Action, pH	Maximum Work Solution Conc. Recommended, %	Maximum Conc. Recommended for Potable Water Use (by US PHS), mg/l	Manufacturer and/or Distributor (Table 4)
							Solution, %	Viscosity, cp	Density, g/ml	pH				
1	2	3	4	5	6	7	8	9	10	11	12	13	14	15
AQUAFLOC 412	N	P	Syn	XP			Orig	50	1.25	1.0		1:4 (D)		27
414	N	P	Syn	PS	977		2	660	1	7		0.2 (2)		27
415	A	L		XP			Orig	1,000	1.03	11.5		10:1 (D)		27
418	C	P	Syn	XPSC			2	23	1	2.4		0.1 (1)		27
419	C	P	Syn	XP								<3%		27
421	A	P	Syn	XP								0.1		27
(Separan NP 10) 422	N	P	Syn	PAM									1	27
431	C	L	Syn	PAM			Orig	150	1.15	6–7		D		27
AQUARID 49–702	C	P	Syn	PAN	577					7–9				93
49–700	C	L	Syn	PAN			Orig	100–500	1.1	7–8		10:1 (D)		93
49–701	C	L	Syn	PAN			Orig	25–150	1.1	7–8		10:1 (D)		93
49–703	A	L	Syn	PS			Orig	200–500	1.2	11.5		20:1 (D)		93
ARAFLOC A-185			Syn	PAM										65
ARFLOC														8
ARIPOL F1			Syn	PAM-COA									PWG	11
ARQUAD 2HT	A	L	Syn											7
2TC	A		Syn	PS-N									6	7
ATLASEP 11													1 (PWG)	60
44													1 (PWG)	60
77													1 (PWG)	60
100													1 (PWG)	60
255													1 (PWG)	60
AZYME			Nat	X										44

1 — Flocculant Trade Name	2	3	4	5	6 — Bulk Density, g/l	7 — Molecular Weight, ×10⁶	8 — Solution, %	9 — Viscosity, cp	10 — Density, g/ml	11 — pH	12 — Optimum Range of Action, pH	13 — Maximum Work Solution Conc. Recommended, %	14 — Maximum Conc. Recommended for Potable Water Use (by US PHS), mg/l	15 — Manufacturer and/or Distributor (Table 4)
		— Chemical Composition —						— Water Solution Properties —						
BARBAFLOC BF-115			Nat	CB										
BERDELL-Flocculant N-489	N												1	10
N-821	N												1	10
N-902	N												1	10
BETZ POLYMER (Betz Poly-Floc) 1100P	A	P	Nat	PAM-C	560		0.5	2,000	1	8.5		0.5	1	12
1110P	A	P	Nat	PAM-CO	592		0.5	2,400	1	7.5		0.5	1	12
1120P	A	P	Nat	PAM-C	592		0.5	3,500	1	8.8		0.5	1	12
1130P	A	P	Nat	PAM-CO	624		1	3,000	1	8.8		0.5	1	12
1140P													1	12
1150P	A	P	Nat	PAM-C	576		1	1,000	1	6-7		1	1	12
1160P	C	P	Syn	PAM-CO	560		1	1,300	1	5-6		1	1	12
1170		L		PAM-C			Orig	50	1.06	9				12
1175		L		PS			Orig	700	1.08	9				12
1200P	A	P	Syn	PAM-C									1	12
1205P													1	12
1210P													1	12
1220P													1	12
1230P													1	12
1240P													1	12
1250P													1	12
1260P													1	12
DK-52	C	L	Syn		X									12
BLANOSE														44
BONDFLOC No. 1-101													5	13

Flocculant Trade Name	- - - Chemical Composition - - -				Bulk Density, g/l	Molecular Weight, x 10⁶	Water Solution Properties				Optimum Range of Action, pH	Maximum Work Solution Conc. Recommended, %	Maximum Conc. Recommended for Potable Water Use (by US PHS), mg/l	Manufacturer and/or Distributor (Table 4)
							Solution, %	Viscosity, cp	Density, g/ml	pH				
1	2	3	4	5	6	7	8	9	10	11	12	13	14	15
BRENCO 870													100	16, 17
880													100	16, 17
N-7024	N												1	16, 17
BUDOND 60	C		Nat	S										18
61	C													18
BURTONITE N 78	N	P	Nat	L	592–865		0.5	300	1	5–6		0.5	5	20
BTI A 100	A	P	Syn	PAM-C							WR	0.2–0.5	Type	14, 14a
A 110	A	P	Syn	PAM-CO							ALK	0.2–0.5	PWG	14, 14a
A 130	A	P	Syn	PAM-C							NEU + ALK	0.2–0.5	max.	14, 14a
A 150	A	P	Syn	PAM-C							ALK	0.2–0.5	1	14, 14a
C 100	C	P	Syn	PAM-C							AC	0.2–0.5	PWG	14, 14a
C 110	C	P	Syn	PAM-S							AC	0.2–0.5	PWG	14, 14a
N 100	N		Syn	PAM-S							WR	0.2–0.5	<1 (PWG)	14, 14a
CL 40	C	L	Syn											14, 14a
CL 80	C		Syn											14, 14a
A 12	A	P	Syn									0.25		14, 14a
LP 150	C	L	Syn	PAD										14, 14a
CALGON Coagulant Aid														
2	N	P	Syn	PS	592–752		0.5	210	1	7–8		0.5	1	21
7	N	P											0.5	21
11	N	P											4	21
18	N	P		SM	1,185–1,409		3	80	1	9–10	WR	3	15	21
37	A	P		SM	945–1,073		1.5	240	1.08	9–10		1.5		21
50	N	P	Syn	PS	720–913		2	250	1.08	8.5		2		21
223	C	L	Syn	PS			Orig	330	1.20	11.7		< 1-1(D)	1	21

Flocculant Trade Name	- - - Chemical Composition - - -				Bulk Density, g/l	Molecular Weight, x 10⁶	- - - - Water Solution Properties - - - -				Optimum Range of Action, pH	Maximum Work Solution Conc. Recommended, %	Maximum Conc. Recommended for Potable Water Use (by US PHS), mg/l	Manufacturer and/or Distributor (Table 4)
							Solution, %	Viscosity, cp	Density, g/ml	pH				
1	2	3	4	5	6	7	8	9	10	11	12	13	14	15
CALGON Coagulant Aid														
225 (= Polymer)	C	L	Syn	PS			Orig	2,000	1.12	8.4		>10-1(D)		21, 21b
226 (= Polymer)	C	P	Syn	PS			0.5	20	1.0	4.0				21b
230	N	P	Syn	PS	156–208		0.5	20	1	6–7		0.2-0.5		21
233 (= Polymer)	N	P	Syn	PS	288–352		0.2	10	1	5–7 (7.5)		0.5	1	21, 21b
240	A	P	Syn	PS	192–272		0.25	700–2,000	1	7–8		0.2-0.25		21
243 (= Polymer)	A	P	Syn	PS	512–592		0.2	200–750	1	7.5		0.25		59
252	A	P	Syn	PS	352–432		0.2	100–250	1	7.5		0.25		21
253 (= Polymer)	A	P	Syn	PS	384–464		0.2	110–300	1	6–7.5		0.25	1	21, 21b 59
ST 260	C	P												21
266	C	L	Syn	PS	448–528		Orig	2,000	1.12	8.4		>10-1(D)		21
269	A	P	Syn	PS			0.2	200–750	1.1	7–9		0.25		21
801													6	21
952													8	21
961	N	P	Syn	PS	416–512		1	7–10	1	5.7–7		10	5	21
C-2256	C	P	Syn	XP	544		1	75	1	6.4		1.5		21
C-2260	C	P	Syn	XP	448		1	40	1	6.7		1.5		21
C-2270	C	P	Syn	XP	400		1	35	1	7.3		1.5		21
C-2300	N	P	Syn	XP	208		0.5	24	1	7.0		0.5		21
C-2325	A	P	Syn	XP	208		0.5	275	1	7.0		0.5		21
C-2350	A	P	Syn	XP	224		0.25	425	1	7.0		0.25		21
C-2400	A	P	Syn	XP	208		0.25	740	1	7.0		0.25		21
C-2425	A	P	Syn	XP	352		0.25	800	1	7.0		0.25		21
WT-2600	C	P	Syn	XP	432		1	38	1	4		1.5		21
WT-2630	C	P	Syn	XP	528		1.5	20	1	4		1.5		21
WT-2635 (= Polymer)	C	L	Syn								WR			21

Flocculant Trade Name	--- Chemical Composition ---				Bulk Density, g/l	Molecular Weight, x 10^6	---- Water Solution Properties ----				Optimum Range of Action, pH	Maximum Work Solution Conc. Recommended, %	Maximum Conc. Recommended for Potable Water Use (by US PHS), mg/l	Manufacturer and/or Distributor (Table 4)
							Solution, %	Viscosity, cp	Density, g/ml	pH				
1	2	3	4	5	6	7	8	9	10	11	12	13	14	15
CALGON Coagulant Aid WT-2640 (= Polymer)	C	L									WR			21
WT-2660 (St-260)	C	P	Syn	XP	496		0.5	20	1	4		0.5		21
WT-2690	N	P	Syn	XP	192		0.5	23	1	7		0.5		21
WT-2700	A	P	Syn	XP	224		0.25	80	1	7.5	WR	0.25		21
WT-2900	A	P	Syn	XP	160		0.25	160	1	7.5		0.25		21
WT-3000	A	P	Syn	XP	350		0.25	250	1	7.5		0.25		21
CARAFLOK 91 AP	A	P	Syn	PAM							4–12 (8)			114
95 AP	A	P	Syn	PAM							4–12 (8)			114
CARBIDE POLYMERS (16 types)														
CAT-FLOC (= WT-2870)	C	L	Syn	DAMA		~0.5	Orig	2,000	1.022	~4.2	WR	<10:1	<7	21a, 21b, 21c
B				DAMA-M									10	21a, 21b
CELLIN			Nat	CP										44
CELLULOSE-GUM	A		Nat	CP										53
CELLUFIXA			Nat	CP										104
CERON CN	C	P	Nat	(CP)										53
CIRRASOL Z														60a
CLARIFIER No. 1		L												87
CLARON		L		BIC			Orig	130	1.45	11–12		<1–10	1.5	2
207		L		BIC			Orig	100	1.185	11–12		<1–10	2	2
CMC 12H	A	P	Syn	CP	672–801		1	110–300	1	7–8				53

Flocculant Trade Name	- - - Chemical Composition - - -				Bulk Density, g/l	Molecular Weight, $\times 10^6$	- - - - Water Solution Properties - - - -				Optimum Range of Action, pH	Maximum Work Solution Conc. Recommended, %	Maximum Conc. Recommended for Potable Water Use (by US PHS), mg/l	Manufacturer and/or Distributor (Table 4)
							Solution, %	Viscosity, cp	Density, g/ml	pH				
1	2	3	4	5	6	7	8	9	10	11	12	13	14	15
CMC 120H	A	P	Nat	CP										53
H-34-52	AC		Nat	CP										53
CARBOXYMETHYL-CELLULOSE													1	33, 53
COAGULANT AID 70		L	Nat				Orig	35–600	1	12		D	1 (PWG)	46
72A	A	L		SM-AN			Orig	65	1.32	11.5		D	50	46
74B	N	L	Syn	PS			Orig	4,500–13,000	1	4–5		D	30	46
76	A	L	Syn	PS			Orig	4,000–16,000		1	4–5	D	40	46
76A													50	46
788	C	L	Syn	PS			Orig	4,000–16,000		1	3.5–4.5	D	50	46
CYANAMER A370	A	P	Syn	PAM										25
P250	N	P	Syn	PAM										25
P260	A	P	Syn	PAM										25
CYANAMID 101			Syn											5
201			Syn											5
301			Syn											5
S-3409														5
3037														5
CW														94
DANSIL K645			Nat	S										94
DC 23		P				0.1–0.2	0.25			4.1				62
DEALCO CI														22
DECABLOC A1	A	B	Syn	PAM										43a

Flocculant Trade Name	--- Chemical Composition ---				Bulk Density, g/l	Molecular Weight, $\times 10^6$	Water Solution Properties				Optimum Range of Action, pH	Maximum Work Solution Conc. Recommended, %	Maximum Conc. Recommended for Potable Water Use (by US PHS), mg/l	Manufacturer and/or Distributor (Table 4)
							Solution, %	Viscosity, cp	Density, g/ml	pH				
1	2	3	4	5	6	7	8	9	10	11	12	13	14	15
DECABLOC A2	A	B	Syn	PAM										43a
A3	A	B	Syn	PAM										43a
A4	A	B	Syn	PAM										43a
A5	A	B	Syn	PAM										43a
N1	N	B	Syn	PAM										43a
C1	C	B	Syn	PAM										43a
C2	C	B	Syn	PAM										43a
DECAPOL A 30	A	P	Syn	PAM							WR		<1 (PWG)	42, 43, 43b
A 33	A	P	Syn	PAM							WR		<1 (PWG)	42, 43, 43b
A 39	A	P	Syn	PAM							WR		<1 (PWG)	42, 43, 43b
A 45	A	P	Syn	PAM							WR		<1 (PWG)	42, 43, 43b
A 48	A	P	Syn	PAM							WR			42
A 51	A	P	Syn	PAM							WR			42
A 57	A	P	Syn	PAM							ALK			42
N 100	N	P	Syn	PAM										42
N 100P	N	P	Syn	PAM									PWG	42, 43b
C 100P	C		Syn	PAM									<10	43
C 300	C	P	Syn	PAM									PWG	43b, 42
C 330	C	P	Syn	PAM									PWG	42, 43b
C 420	C	P	Syn	PAM										42
C 510	C	L	Syn	PAM										42
C 570	C	L	Syn	PAM										42
D 600	C	L	Syn	PAN										42
D 610	C	L	Syn	XP										42

Flocculant Trade Name	- - - Chemical Composition - - -				Bulk Density, g/l	Molecular Weight, x 10^6	Water Solution Properties				Optimum Range of Action, pH	Maximum Work Solution Conc. Recommended, %	Maximum Conc. Recommended for Potable Water Use (by US PHS), mg/l	Manufacturer and/or Distributor (Table 4)
							Solution, %	Viscosity, cp	Density, g/ml	pH				
1	2	3	4	5	6	7	8	9	10	11	12	13	14	15
DELFLOC (= HERCOFLOC)														53a
DF 381														94
454	C		Syn								WR			31
DOW A-21	A													28
A-22	A													28
A-23	A													28
N-11	N													28
N-12	N													28
647			Syn											28
1128-D														28
DOWELL M-143													5	28
DRESINATE TX														53
XX														53
DREWFLOC 1	(N)	L	Syn	APC(S)			Orig	100–150	1.47	11–12		>5	PWG	31
3		P	Syn	XP(S)	464–576		2	35–50	1	6–7.5		0.5–2	3	31
4													5	31
5	N	P	Syn	XP										31
21	C	P	Nat	CPN	280–384		2	25–50	1	6–7.5		0.5–2	5	31
31	C	L	Syn	XPSC										31
35	C	L	Syn	PS			Orig	Viscous	1	6–7	WR	<5		31
138	C		Syn	XP										31
139	C		Syn	XP										31
225													1	31
265													1	31
922													10	31

Flocculant Trade Name	Chemical Composition				Bulk Density, g/l	Molecular Weight, $\times 10^6$	Water Solution Properties				Optimum Range of Action, pH	Maximum Work Solution Conc. Recommended, %	Maximum Conc. Recommended for Potable Water Use (by US PHS), mg/l	Manufacturer and/or Distributor (Table 4)
							Solution, %	Viscosity, cp	Density, g/ml	pH				
1	2	3	4	5	6	7	8	9	10	11	12	13	14	15
DREWFLOC Ser. 400	C	L	Syn	XP			1	3.5	1.1	5-6	WR	D		31
DT 120		L	Syn	PA							ALK + AC	<1		36
220		L	Syn	PA							AC	<1		36
DUOMAC T	C		Syn											33
DU PONT EXR 102A														
DX 908														78
DYNA FLOC 631													35.0	34
632													1.0	34
633													1.0	34
634													1.0	34
661													1.0	34
662													5.0	34
664													5.0	34
691													1.0	34
692													1.0	34
693													5.0	34
ECCO Suspension Catalyzer No. 146													3.5	30, 86
EKOFAN														29
ENPROX 607														87
EZ 494													0.5	35
FABCON													0.5	35

Flocculant Trade Name	- - - Chemical Composition - - -				Bulk Density, g/l	Molecular Weight, x 10^6	- - - - Water Solution Properties - - - -				Optimum Range of Action, pH	Maximum Work Solution Conc. Recommended, %	Maximum Conc. Recommended for Potable Water Use (by US PHS), mg/l	Manufacturer and/or Distributor (Table 4)
							Solution, %	Viscosity, cp	Density, g/ml	pH				
1	2	3	4	5	6	7	8	9	10	11	12	13	14	15
FI-CLOR 605			Syn	PAN									<15 (Swimming pools)	39
91			Syn	PAN									<7 (Swimming pools)	39
FILTAFLOK A1	A(N)	P	Syn	PA-BC	~860		~2				4–12 (8)	0.1		106, 112
FLOC-ACE														52
FLOC AID 1038	N	P	Nat	S	720–881		0.5	20–30	1	<6–7		D (2)	5	84
1063	C	P	Nat	S	368–464		0.5	45	1	<6–7		2	5	84
FLOCAL F601														44
T214														44
FLOCBEL 170H			Syn											40
160FG			Syn	PAM-COA									<1 (PWG)	40
FLOCCOTAN M	AC	P or (L)	Nat	Mi									<10	41
A		L	Nat	T		>0.05	2.9			3.0			<10	41
CS 7		P				8	0.26			6.7				41
FLOCCULITE 550													2	32
FLOCGEL 80		P	Nat	S										40
81		P	Nat	S										40
FC40		L	Syn	PN										40
FC167		P												40

Flocculant Trade Name	--- Chemical Composition ---				Bulk Density, g/l	Molecular Weight, $\times 10^6$	---- Water Solution Properties ----				Optimum Range of Action, pH	Maximum Work Solution Conc. Recommended, %	Maximum Conc. Recommended for Potable Water Use (by US PHS), mg/l	Manufacturer and/or Distributor (Table 4)
							Solution, %	Viscosity, cp	Density, g/ml	pH				
1	2	3	4	5	6	7	8	9	10	11	12	13	14	15
FLOCGEL FC187		P	Syn	PAN										40
433														40
1297	C		Syn	PAM-CO										40, 96
FLOCONIT			Syn	PAM										85
FLOCULES CMC			Nat	CP										44
FLOKAL 101		P	Syn	PN										26
202,5		P	Nat	S										26
PAM		P	Syn	PAM										26
CMC		P	Nat	CP										26
301			Syn	PA										26
FORMULA 70A													50	46
73													50	46
74E													20	46
FOXAMID C086			Syn											
GALACTASOL	N	P	Nat	G										97
GAMAFLOC NI-702													4	45
GAMLEN (Wisprofloc 20)													5	45
GAMLOSE W	A		Nat	S-CM								0.5–2.0	5	45, 45a
GELFLOK	N	L	Nat	S-C								D		112
GF 2	C		Syn											98
GIGAMAL P-26			Nat	S-C										92

Flocculant Trade Name	Chemical Composition				Bulk Density, g/l	Molecular Weight, $\times 10^6$	Water Solution Properties				Optimum Range of Action, pH	Maximum Work Solution Conc. Recommended, %	Maximum Conc. Recommended for Potable Water Use (by US PHS), mg/l	Manufacturer and/or Distributor (Table 4)
							Solution, %	Viscosity, cp	Density, g/ml	pH				
1	2	3	4	5	6	7	8	9	10	11	12	13	14	15
GIGAMID			Syn	PAM										92
GIGTAR			Syn	PAM										92
GIPAN			Syn	PN-H										92
GOODRITE K-702		L	Syn	PA			Orig	500–1,500	1.09	2–3	ALK + NEU	D		50
K-705		L	Syn	PA			Orig	1,000–2,000	1.10	7–9		D		50
K-708		L	Syn	PA			Orig	1,500–3,500	1.14	8–10		D		50
K-714		L	Syn	PA			Orig	3,500–15,000	1.05	2–3		D		50
K-716		L	Syn	PA			Orig	3,000–15,000	1.04	7–9		D		50
K-718		L	Syn	PA			Orig	3,000–15,000	1.03	7–9		D		50
GUARTEC GC 1	N	P	Nat	G	624–817		1	300–2,500	1	6–7		1		48
SS 1	N	P	Nat	G	624–849		1	2,500–4,500	1	6–7		1		48
HAGAN Coagulant Aid (see CALGON Coagulant Aid)														
HALLMARK 81	N	P	Nat	G	416–512		1	13	1	8.9		2	1	101
82													1	101
HAMACO 196		P	Nat	S	336		1	<50	1	6–7		2	5	100
HERCOFLOC 800 (= RETEN 205)	C	P	Syn	PA	400–500	1.0	1	700–1,000	1	5–6			0.5	53, 53a
810	C	P	Syn	PA	600–700	1.5	1	1,300–1,800		5–6				53, 53a
814	C	P	Syn	PA	700–800	1	1	200–500		5–6			0.5	53, 53a
816	A	P	Syn	PA	600–700	2	1	1,000–1,500		7–8				53, 53a
818	A				528		0.5	1,000	1	8–9		0.5	1 (PWG)	53, 53a
819	A	P	Syn	PA	700–800	8–10	1	3,500–6,000		7–8			1	53, 53a

| Flocculant Trade Name | Chemical Composition | | | | Bulk Density, g/l | Molecular Weight, x 10^6 | Water Solution Properties | | | | Optimum Range of Action, pH | Maximum Work Solution Conc. Recommended, % | Maximum Conc. Recommended for Potable Water Use (by US PHS), mg/l | Manufacturer and/or Distributor (Table 4) |
| | | | | | | | Solution, % | Viscosity, cp | Density, g/ml | pH | | | | |
1	2	3	4	5	6	7	8	9	10	11	12	13	14	15
HERCOFLOC 821													1 (PWG)	53, 53a
824	N	P	Syn	PA	400–500	2	1	300–500		7–8				53, 53a
829	C		Syn											53, 53a
831	A	P	Syn	PA	700–800	6–8	1	3,500–4,000		7–8				53, 53a
HX 19														44
HYPAN			Syn	PAM-CON										92
CHITOSAN			Nat	APD										
ICS 101	C	L	Syn	X										63
162	A	L	Syn	X										63
ILLCO IFA 313		L	Syn	PS			Orig	15,000	1.08	6.6–7.5		>4–1 (D)	10	61
IM 100		P	Syn	PMA										98
IONAC Coagulant Aid 75	C	L	Nat	S			Orig	700–1,000	1.04	8–9		D	5	64, 91
PE-100	C	L (50%)	Syn	PAN					1.16	7				64, 91
NI-702													4	64, 91
NA-710	A	P	Syn	PX	640		0.2	150	1	5–6		1		64
Wisprofloc 20	N	P	Nat	S	288–384		2	80	1	8.9		2	5	64, 91
Wisprofloc 75													5	64, 91
JAGUAR (Ser.) 500		P	Nat	G								0.05–0.1		76, 101
703		P	Nat	G										76, 101
705		P	Nat	G										76, 101
800	A	P	Nat	G										76, 101
J.2.S.1.	N	P	Nat	G										76, 101

Flocculant Trade Name (1)	Chemical Composition (2)	(3)	(4)	(5)	Bulk Density, g/l (6)	Molecular Weight, $\times 10^6$ (7)	Solution, % (8)	Viscosity, cp (9)	Density, g/ml (10)	pH (11)	Optimum Range of Action, pH (12)	Maximum Work Solution Conc. Recommended, % (13)	Maximum Conc. Recommended for Potable Water Use (by US PHS), mg/l (14)	Manufacturer and/or Distributor (Table 4) (15)
JAGUAR Nonionic	N	P	Nat	G			~1	~3,300						76, 101
MD-7A		P	Nat	G										76, 101
MDB		P	Nat	G										76, 101
MDC		P	Nat	G										76, 101
MDD	N	P	Nat	G	728–810		1	2,800–3,200		5.9–6.3	0.5–12	0.05–0.5		76, 101
MDS	N	P	Nat	G										76, 101
PLUS	C	P	Nat	G	350		1	800	1	8.0–9		1		76, 101
WP-B	N	P	Nat	G	608–688		0.25	10	1	5–6		0.25	0.5	76, 101
WP-D	N	P	Nat	G	576–768		0.25	10	1	5–6		0.25	0.5	76, 101
JM 67			Nat	S									<2	66
JAYFLOC 67			Nat	S									<2	66
K-4			Syn	PN-H										98
K-6			Syn	PN										98
K-7				(PO_4)										98
KATAFLOC	C	L	Nat	KO										72
KELCOSOL		P	Nat	AL	320–496		0.5	350	1	7–8		0.5	2 (5)	67, 68
KELGIN W		P	Nat	AL	640–784		0.5	240	1	7–8		0.5	2 (5)	67, 68
KELTEX		P												67
KESTREL 220			Syn	PAM										
221			Syn	PAM										
KEY-FLOC (4% water solution of SEPARAN NP-10) W	N	L (4%)	Syn	PAM			4.0	1,100	1.1	6–7		D	25	69

94

Flocculant Trade Name	Chemical Composition				Bulk Density, g/l	Molecular Weight, x 10^6	Water Solution Properties				Optimum Range of Action, pH	Maximum Work Solution Conc. Recommended, %	Maximum Conc. Recommended for Potable Water Use (by US PHS), mg/l	Manufacturer and/or Distributor (Table 4)
							Solution, %	Viscosity, cp	Density, g/ml	pH				
1	2	3	4	5	6	7	8	9	10	11	12	13	14	15
KLEER-FLOC 102													1	38
109													1	38
116													5	38
119													1	38
730													1	38
735													1	38
KOAZOL 399	C		Nat	S										
KODT	C		Syn	KODT										98
KOMETA			Syn	PAM										98
KRILIUM (= VAMA) 6		P	Syn	VAM										78, 79
9		P	Syn	PA										78, 79
KURIFLOC			Syn	PA-AA										70
LACCO Polymer Coagulant														
LYTRON 820	A	P	Syn	VAM										78, 79
825	A	P	Syn	VAM										78, 79
886	A	P	Syn	VAM	640–849		2	25	1	3–5	7–8	1–2		78, 79
887	A	P	Syn	VAM										78, 79
LURESIN K	C	L	Syn	PA			Orig	<100	1.03	4–5	AC + NEU			15
PR 8035	C	L	Syn	XP			Orig	200–400	1.05	3.5–4.5	AC + NEU			15
LUSTREX (= Lytron) X-886	A	P	Syn	VAM										78
X-889														78
MAFLOC Syn	C	L		AP			Orig			1.2–1.25	2.5–3		1:20 (D)	74

| Flocculant Trade Name | - - - Chemical Composition - - - | | | | Bulk Density, g/l | Molecular Weight, x 10⁶ | - - - - Water Solution Properties - - - - | | | | Optimum Range of Action, pH | Maximum Work Solution Conc. Recommended, % | Maximum Conc. Recommended for Potable Water Use (by US PHS), mg/l | Manufacturer and/or Distributor (Table 4) |
| | | | | | | | Solution, % | Viscosity, cp | Density, g/ml | pH | | | | |
1	2	3	4	5	6	7	8	9	10	11	12	13	14	15
MAGNAFLOC E 24	A	P												3
HR-120	C	L	Syn	PAM		1	0.76			5.1	ER	0.1		3
LT 20			Syn	PAM-COA									PWG	3
Type LT for potable water treatment LT 22	C		Syn										1	3
LT 24	C	P	Syn	PAM-S		>2						0.01–0.05	1	3
LT 25	A	P	Syn	PAM-S		>2						0.01–0.05	PWG	3
LT 26			Syn										PWG	3
R-139	A	L	Syn	PAM							3–11	0.1		3
R-140	C	P	Syn	PAM							ER	0.1		3
R-155	A	P	Syn	PAM							3–11	0.1		3
R-156	A	P	Syn	PAM							4–10	0.1		3
R-159														3
R-225	C		Syn	PAM-S										3
R-292	C	P	Syn	PAM		5	0.23			~6				3
351	N	P										0.1		3
352	C	P										0.1		3
455	C	P										0.1		3
520	C	P												3
EM-127	A	L	Syn	PAM							5–10			3
(low MW EM-127) EM-15	A	L	Syn	PAM							5–10			3
MAGNIFLOC (=Superfloc) 521-C	C	L	Syn	PS			Orig	225–325	1.15	4–5		<10:1 (D)	10	5
(Ser. 500-cationic) 530-C	C	P	Syn	PX								0.5		5, 25
570-C	C												10	5
571-C													10	5
(= Superfloc) 573-C	C												10	5

Flocculant Trade Name	- - - Chemical Composition - - -				Bulk Density, g/l	Molecular Weight, x 10⁶	- - - - Water Solution Properties - - - -				Optimum Range of Action, pH	Maximum Work Solution Conc. Recommended, %	Maximum Conc. Recommended for Potable Water Use (by US PHS), mg/l	Manufacturer and/or Distributor (Table 4)
							Solution, %	Viscosity, cp	Density, g/ml	pH				
1	2	3	4	5	6	7	8	9	10	11	12	13	14	15
MAGNIFLOC 575-C	C												10	5
(= Superfloc) 577-C	C												10	5
579-C	C												10	5
(= Superfloc) 581-C	C	L	Syn										10	5
845-A	A												1	5
(Ser. 800-anionic) 846-A	A												1	5
847-A	A												1	5
860-A	A	P	Syn	PAM	640–768		1	68–78	1	4–5		1	≤1	5, 25
(= Superfloc) 971N	N	P	Syn	PAM	416–528		1	115–200	1	<6–7		1	1	5, 25
(Ser. 900-nonionic) 972N	N	P	Syn	PAM	448–544		1	160	1	<6–7		1	1	5, 25
985N	N	P	Syn	PAM	448–528		1	250–750	1	6–7		1	1	5, 25
990	N	P	Syn	PAM	432–512	4	1	180	1	4–5		1	1	5, 25
METALLENE Coagulant P-6	C	L	Syn				Orig	50–70	1.45	14		D	5	75
METAS			Syn											98
METHAMYL (Na, NH₂)														
MEYPRO-QUAR														
MOGUL-CLARACEL (= Claron 207) CO-980		L		BIC			Orig	100	1.185	11–12		<1–10	2	86
CO-981		L	Syn	PS			Orig		1.04	8.3		D		86
(= Claron) CO-982		L	Syn	PAM			Orig	130	1.45	11–12		<1–10	1.5	86
(= Separan NP 10) CO-983	A	L	Syn	PAM			Orig	3,000–4,000	1	6–7		>10–1 (D)	1	86
CO-984		L	Syn	PS			Orig		1.003	9.8		D		86
CO-985													3.5	86
CO-986													5	86
NALCO 600	C	L	Syn	PAN-H			Orig	40	1.078	11.2		>10–1 (D)		80, 81

Flocculant Trade Name	- - - Chemical Composition - - -				Bulk Density, g/l	Molecular Weight, x 10^6	- - - - Water Solution Properties - - - -				Optimum Range of Action, pH	Maximum Work Solution Conc. Recommended, %	Maximum Conc. Recommended for Potable Water Use (by US PHS), mg/l	Manufacturer and/or Distributor (Table 4)
							Solution, %	Viscosity, cp	Density, g/ml	pH				
1	2	3	4	5	6	7	8	9	10	11	12	13	14	15
NALCO (Spec.) 600	C	L	Syn	XP			Orig	700–4,000	1.14	12.4				80, 81
610	C	P												80, 81
633	A	P	Syn	PS	608–736		0.5	600	1	6.8		$<$1–1.5		80, 81
633-HD	C	P	Syn		836		2.0	190	1	8.0				80, 81
634	C	L	Syn	XP				70–250	1.175	9.5				80, 81
635	A	P	Syn	XP	720		0.25	2,500	1	8.5		$<$0.25		80, 81
636	C	P	Syn	XP	760		0.5			4.7		0.25–0.75		80, 81
636HD	C	P	Syn		752		1.5	33	1			1.5		80, 81
650	A	P		SM	849–1,105		6	40–360	1.03	9–10		6		80, 81
D-1444	AC	L												80, 81
D-2339	A	P	Syn	PS	688		0.3	2,000	1	7.0		0.5		80, 81
NALCOLYTE 12C06												0.5–1.0	10	80, 81
110	N	P	Nat	S	208–272	$>$1	5	60	1	6–7	$<$5	15	5	80, 81
603	C	L	Syn	PS	956		Orig	50–100	1.177	9.3	WR	10 (D) (1:10, 1:100)		80, 81
605	C	L	Syn	PS			Orig	70	1.084	11–12	WR			80, 81
607	C	L	Syn	XP	968		Orig	50	1.17	7.5		$<$1:10 (D)	7	80, 81
610		P	Syn	XP	820	$>$1	1	3,000		5.1		$<$1		80, 81
670	N	P	Syn	PAM	544–720	$>$1	1	50–200	1	7–8		$<$1		80, 81
671	N	P	Syn	PS	528–656		1	450	1	7–8		0.02–1	1	80, 81
672	A	P	Syn	PAM	560–768	$>$1	0.5	150	1	6–7		0.01–0.5		80, 81
673	A	P	Syn	PS	672–784	$>$1	0.3 (1.0)	2,500	1	(8.5)		0.01–0.3		80, 81
675	A	P	Syn	PS	820	$<$1	0.25 (1.0)	2,500	1	(8.5)		0.25		80, 81
677														80, 81

Flocculant Trade Name	- - - Chemical Composition - - -				Bulk Density, g/l	Molecular Weight, x 10⁶	Water Solution Properties				Optimum Range of Action, pH	Maximum Work Solution Conc. Recommended, %	Maximum Conc. Recommended for Potable Water Use (by US PHS), mg/l	Manufacturer and/or Distributor (Table 4)
							Solution, %	Viscosity, cp	Density, g/ml	pH				
1	2	3	4	5	6	7	8	9	10	11	12	13	14	15
NALCOLYTE 7870													1	80, 81
8101	C												10	80, 81
8113													10	80, 81
8114													10	80, 81
8170													1	80, 81
8171													1	80, 81
8172		L	Syn								5-10		1	80, 81
8173													1	80, 81
8174			Syn								7-14			80
8175													1	80, 81
D-1702	A	P	Syn	PAM										80
D-1782	A	P	Syn	PAM							ALK			80
NALFLOC (= Nalco products)														82
A 373	A		Syn	PAM-COA									<1	82
A 375	A		Syn	PAM-COA									<1	82
A 378	A		Syn	PAM-COA									<1	82
605	C		Syn											82
N 607	N		Syn	PAN									<7	82
N 671	N		Syn										<1	82
NATRON 86	C	L	Syn	PS			Orig	300-500	1.06	3.0	WR	<0.5		84
6082	AC		Syn											84
NATROSOL 250HR	N	P	Nat	HEC			2	25,000						53
NEOGUM			Nat	XP										
O'B—FLOC		P	Nat	S	384-512		2	10	9-10			2	10	88

Flocculant Trade Name	Chemical Composition				Bulk Density, g/l	Molecular Weight, $\times 10^6$	Water Solution Properties				Optimum Range of Action, pH	Maximum Work Solution Conc. Recommended, %	Maximum Conc. Recommended for Potable Water Use (by US PHS), mg/l	Manufacturer and/or Distributor (Table 4)
							Solution, %	Viscosity, cp	Density, g/ml	pH				
1	2	3	4	5	6	7	8	9	10	11	12	13	14	15
ORZAN L50														24
P		P	Nat	CH										24
SL45														24
OXFORD-HYDROFLOC													10	89
P-25			Nat	S										92
PAA 3S		P	Syn	PAM										
SGS		P	Syn	PAM										
SGK		P	Syn	PAM										
SSPA		P	Syn	PAM										
MF		L (0.5%)	Syn	PAM										
PAM 50	A		Syn	PAM										5
75	A		Syn	PAM										5
100	A		Syn	PAM										5
PANG			Syn	PN										98
PEI 6	C		Syn	PEI		0.0006								28
12	C		Syn	PEI		0.0012								28
18	C		Syn	PEI		0.0018								28
600	C		Syn	PEI		0.04–0.06								28
1000	C		Syn	PEI		0.1								28
1090	C		Syn	PEI									5	28
1120	C	P												28
PERCOL LT-20													1	3b
LT-22													1	3b
LT-24													5	3a, 3b

Flocculant Trade Name	Chemical Composition				Bulk De Bulk Density, g/l	Molecular Weight, x 10⁶	Water Solution Properties				Optimum Range of Action, pH	Maximum Work Solution Conc. Recommended, %	Maximum Conc. Recommended for Potable Water Use (by US PHS), mg/l	Manufacturer and/or Distributor (Table 4)
							Solution, %	Viscosity, cp	Density, g/ml	pH				
1	2	3	4	5	6	7	8	9	10	11	12	13	14	15
PERCOL LT-25													5	3a, 3b
LT-26													1	3a, 3b
LT-29													1	3b
PERFECTAMYL A5114/2			Nat	S									<2	54, 55
PERMUTIT 65	AC	P											2	107
66													2	107
67													4	107
(= Wisprofloc 20) 68													5	107
PMA-848			Syn	MA										105
POLIMIN														15
FL	C		Syn	PAM										15
P	C	L (50%)	Syn	PEI			Orig	10,000–2,000	1.07		5.5–8			15
POLYAKRYLAMID	A	L (8%)	Syn	PAM			1	6.8		7	4–8	D		98
POLYAKRYLNITRIL	A		Syn	PN										
POLYFLOK A1	A										2			106, 112
(4% water solution of Separan NP-10) 4D	A	L (20%)	Syn	PA (Na)		~0.05	4	1,100	1.1	6–7	5–14 (8)	2 (D) (1:50)5	25	3, 12, 106
37cP	C	L (15%)	Syn	PA-A		~1	Orig		1.03	9	3–8	1% (0.5)		3, 106
100X	A	L (6%)	Syn	PA							6–9			3, 106
209P	N	P	Syn	PAM-AN	~470	>10				ALK	2–12 (8)	0.2		3, 106
365P	A (N)	P	Syn	PAM-AN		~6						0.2	<1	3, 106
430P	A (N)	P	Syn	XP+AN		~2					6–10	0.1–0.2	<1	3, 106
520P	A (N)													

Flocculant Trade Name	Chemical Composition				Bulk Density, g/l	Molecular Weight, x 10⁶	Water Solution Properties				Optimum Range of Action, pH	Maximum Work Solution Conc. Recommended, %	Maximum Conc. Recommended for Potable Water Use (by US PHS), mg/l	Manufacturer and/or Distributor (Table 4)
							Solution, %	Viscosity, cp	Density, g/ml	pH				
1	2	3	4	5	6	7	8	9	10	11	12	13	14	15
POLYFLOK 856P	A (N)	P	Syn	PA-BC	~566	>10					4–12 (8)	<1%		3, 106
8H	A	L	Syn	PA							6–9			3
LP	A	L	Syn	PAM										3
PX	A (N)	L (10%)	Syn	PA-B		~2	Orig		1.02	7–8	2–12 (5–9)	2		3
XL	A	L	Syn	PAM										3
PR-8039	A	L	Syn	PA			Orig	3,000–6,000		4–5	AC			15
8079														15
POLY-FLOC (see Betz Polymer)														
POLYHALL M-245	A	P	Syn	PAM	400–512		1	4,500±	1	7.0±		0.5	PWG	101
M-295 W.P.	A	P	Syn	HPAM	~470		1	4,500±		7.0±	5.5–12	<0.5	1	101
402	N	P	Syn	PAM	400–485		1	400–900		5–6	0.5–11.5	<0.5		101
M-630	C	L	Syn	PX			Orig	200–500	1.05	4±0.5	0.5–9.0	<1.0		101
M-19	N		Syn	PAM										101
POLYMER F-3	A	P	Nat	G	544		4	800	1	9.5		4		101
(Polymer 225, 226, 233, 243, 253—see Calgon Aid)														
7050	N	P	Nat	X										71
X-150	C	L	Syn	PS			Orig	2,000–6,000	1.08	3.5–5.5		1		71, 108
X-1633														71, 108
POLYMIN HS	C	L (20%)	Syn	PEI			Orig	500–1,000	1.07	7–8	5–7.5			15
KM	C	L (20%)	Syn	PEI			Orig	500–1,500	1.068	~8	4–8			15
PR	A	P	Syn	PAM	700							0.1–1		15
SN	C	L (20%)	Syn	PEI			Orig	800–1,800	1.06	7.5–8.5	4–7.5			15
POLYOX	N	P	Syn	PAN	400–512	>4	1	5,500	0.98	9–10		1		108
Coagulant	N	P	Syn			>4								108

Flocculant Trade Name	--- Chemical Composition ---				Bulk Density, g/l	Molecular Weight, x 10^6	Solution, %	Viscosity, cp	Density, g/ml	pH	Optimum Range of Action, pH	Maximum Work Solution Conc. Recommended, %	Maximum Conc. Recommended for Potable Water Use (by US PHS), mg/l	Manufacturer and/or Distributor (Table 4)
1	2	3	4	5	6	7	8	9	10	11	12	13	14	15
POLYOX MN (WSR-301)	N	P	Syn	PEN		>4								108
POLYTERIC AS 3			Syn	HPAM										108
POWDAFLOK	N (A)	P	Nat	S-C							~10	~5		3, 112
PRAESTOL FZ	N	L (9%)	Syn	PN										57
103	A	L (11%)	Syn	PAM										57
114	A (N)	L (10%)	Syn	PN	270-330	<10					5-12 (6-10)	(1:1,500) D		57
A114	N (A)	P	Syn	PAM-CO	200-280	5-10					3-12 (5-9)			57
(= 2900) 115	A	L (10%)	Syn	PN										57
130		L (7%)	Syn	PAM										57
135	A		Syn											
215		L (10%)	Syn	PN										57
229	A		Syn											57
2115		L (10%)	Syn	PN										57
2290		L (10%)	Syn	PN										57
2280		L (6%)	Syn	PMA										57
2340		L (6%)	Syn	PMA										57
2400		L (10%)	Syn	PN										57
2450	A (N)	L, Pa	Syn	PA	280-340	1-5					6-12 (6-10)	0.1-1		57
2700	N	L, P	Syn	PAM-CO	215-265	10-15					1-12 (3-6)	0.02-0.1		57
2700 conc.	N	L	Syn	PAM	320-360	<15					1-12 (5-10)	(1:2,000) D		57
2750	N (A)	P	Syn	PAM-CO	230-290	10-15					5-12 (6-9)	(1:1,500) D		57

Flocculant Trade Name	--- Chemical Composition ---				Bulk Density, g/l	Molecular Weight, $\times 10^6$	---- Water Solution Properties ----				Optimum Range of Action, pH	Maximum Work So-lution Conc. Recommended, %	Maximum Conc. Rec-ommended for Po-table Water Use (by US PHS), mg/l	Manufacturer and/or Distributor (Table 4)
							Solution, %	Viscosity, cp	Density, g/ml	pH				
1	2	3	4	5	6	7	8	9	10	11	12	13	14	15
PRAESTOL 2800	N	L (10%)	Syn	PAM	175–225	<3					0–13 (2–6)	0.05–0.1		57
2800	N	P	Syn	PAM	165–195	5–10					0–13 (0–3)			57
2820	N	P	Syn	PAM-CO	175–225	<10					1–12 (2–6)	(1:1,200) D		57
2830	N	P	Syn	PAM-CO	175–225	<10					1–12 (3–7)	(1:1,200) D		57
2850	A (N)	P, L	Syn	PAM-CO	260–320	<3					5–12 (6–10)	0.05–0.5		57
(= A 114) 2900	N	P	Syn	PAM-CO	175–225	<10					1–12 (5–9)	0.05–0.5		57
2935	N (A)	P	Syn	PAM-CO	175–225	<10					5–13 (6–10)	(1:1,500) D		57
3000	N	P	Syn	PAM	175–225	<10					0–13 (0–3)	(1:1,000) D		57
FZS	N	P	Syn	X(KH)	390–410	<0.5					0–13			57
184-K	C	L (20%)	Syn	PAN		0.5					1–13 (4–8)	(1:50) D		57
185-K	C	L (30%)	Syn	PAC		0.3–0.4					1–13 (4–8)	(1:100) D		57
111-K	C	L	Syn	PAN		0.5					4–12 (5–9)			57
222-K	C	L	Syn	PAC	500–560	0.3–0.4					1–13 (4–8)	(1:500) D		57
444-K	C	P	Syn	PAC	550–650	1–3					1–13 (1–8)	(1:500) D		57
500-K	C	P	Syn	PAC	175–225	1–3					4–13 (4–8)	(1:1,000) D		57
PRIMAFLOC C-3	C	L (30%)	Syn	PAN		0.015–0.025	Orig	270	1.2	12–14	ER	0.1–1		71, 77, 94

Flocculant Trade Name	Chemical Composition				Bulk Density, g/l	Molecular Weight, x 10^6	Water Solution Properties				Optimum Range of Action, pH	Maximum Work Solution Conc. Recommended, %	Maximum Conc. Recommended for Potable Water Use (by US PHS), mg/l	Manufacturer and/or Distributor (Table 4)
							Solution, %	Viscosity, cp	Density, g/ml	pH				
1	2	3	4	5	6	7	8	9	10	11	12	13	14	15
PRIMAFLOC C-5	C	L (25%)	Syn	PAN			Orig	1,375	1.14	6–7	ER	D		71, 77, 94
C-6	C	L (20%)	Syn	PA			Orig	300	1.12	12–13		0.5–2		71, 77, 94
C-7	C	P (55%)	Syn	PAN (SO₄)	680–865	>1	2	20	1.01	1.8	3–8	0.5–2 (6)		71, 77, 94
Q-52		L				0.3	2.7			3.5				71
Q-20		L				0.1	1.57			2.9				71
XA-10	A	L (20%)	Syn	PS			Orig	150–250	1.046	~3		0.1–1		71, 77, 94
A-10	A	L (20%)	Syn	XP				150–250	1.046	3		0.1–1		71, 77, 94
PROSEDIM	C													
PURIFLOC A-21	A	P	Syn	SP	592–752		1	600–1,200	1	11–12	WR	1–3		28
A-22	A	P	Syn	PAM	508–736		1	4,800	1	10–11	WR	0.05–1	<1	28, 28a
A-23	A	P	Syn	PAM	720	~8.0	0.1	950	1	9.8		0.5	1 (PWG)	28
C-31	C	L	Syn	PAN		~0.03	Orig	12,000 (5% 10 cp)	1.2	8–10		0.5–5	5	28
C-32	C	L	Syn	PAN		0.15	Orig	20,000	1.2	8–10		0.5–5		28
C-33	C	L	Syn	PEI										28
N-11	N	P	Syn	PAM	432–528		1	490		6–7	WR	0.05–1		28
N-12	N	P	Syn	PAM	496–608		1	880	1	6–7	WR	0.05–1		28
N-17	N	P	Syn	PAM	464–592	8.0	1	490	1	6–7	WR	<1	1 (PWG)	28, 28b
N-20	N												1	28
(= Separan NP 10) NP-10														28
401	N	P	Syn											28
402	N	P	Syn											28
501	A		Syn											28

| Flocculant Trade Name | Chemical Composition | | | | Bulk Density, g/l | Molecular Weight, x 10^6 | Water Solution Properties | | | | Optimum Range of Action, pH | Maximum Work Solution Conc. Recommended, % | Maximum Conc. Recommended for Potable Water Use (by US PHS), mg/l | Manufacturer and/or Distributor (Table 4) |
| | | | | | | | Solution, % | Viscosity, cp | Density, g/ml | pH | | | | |
1	2	3	4	5	6	7	8	9	10	11	12	13	14	15
PURIFLOC 502	A	P	Syn											28
601	C	L	Syn								<7			28
602	C	L	Syn											28
PUSHER 500			Syn	HPAM										
PVM-MA			Syn	MVM										47
QUAR (flour) 22858		P	Nat	G										
QUARTEC F		P	Nat	G									10	48
SJ													10	48
REAGENT (development product) C 177	C		Syn											82
X-309	AC	P												5
S-3019	C	P	Syn	XP										5
(= Aerofloc 3000) S-3000		P	Syn	PAM										5
(= Aerofloc 570) S-3171		P	Syn	PAM										5
182	AC	P	Nat	S										5
REAGENT MRL 13	N	P	Nat	G	400–528		1	13	1	8.6		0.25–2	1	101
14	N	P	Nat	G	464–560		1	13	1	8.8		0.25–2	1	101
19	N	P	Nat	G	120–186		1	20	1	6–7		0.25–1	1	101
(= Jaguar) 22	N	P	Nat	G									1	76, 101
22A	C	P	Nat	G	672–817		1	800–1,200	1	5.5–6.0	4–9	1	1	76, 101
91	C	P	Nat	S-S						8.5–9		<1.0		101
93		P										0.25–2		101
95		P										0.25–2		101
159		P	Syn	XP								0.25–1		101
273														101

Flocculant Trade Name	--- Chemical Composition ---				Bulk Density, g/l	Molecular Weight, x 10⁶	Water Solution Properties				Optimum Range of Action, pH	Maximum Work Solution Conc. Recommended, %	Maximum Conc. Recommended for Potable Water Use (by US PHS), mg/l	Manufacturer and/or Distributor (Table 4)
							Solution, %	Viscosity, cp	Density, g/ml	pH				
1	2	3	4	5	6	7	8	9	10	11	12	13	14	15
REAGENT MRL 295		P	Nat									0.1–0.15		101
364		P	Syn	XP								0.1–0.15		101
364		P										0.25–2		101
601		P										0.05		101
602		P										0.05		101
REAGENT RD (development product) 2180	C		Syn											106a
2181	C		Syn											106a
4053														78
4054														78
4055														78
RELATIN 274		P	Syn	CP										
626		P	Syn	CP										
RESYN 3285	A		Syn								WR			84
RETEN 205(M) (= HERCOFLOC 800)	C	P	Syn	PS	700–750		1	325		6.5	6–10	1		53
205(MH)	C	P	Syn	PS	700		1	840		6.5	6–10			53
205(L)	C	P	Syn	PS	700		1	15		6.5	6–10			53
205-MS	A	P	Syn	XP	592–672		0.5	300–400	1	5–7		0.5		53
210	C	P	Syn	XP			1	1,000		6.5				53
A-M	A	P	Syn	PS	672		1	1,400	1	6.5		1		53
A 1	A	P	Syn	PS	528–704		1	1,400	1	6.5			1	53
A 5	A	P	Syn	PS	624–768		0.25	500–2,500	1	8–9		0.25	1	53
ROHAFLOC L 1			Syn											94
L 2			Syn											94

| Flocculant Trade Name | - - - Chemical Composition - - - | | | | Bulk Density, g/l | Molecular Weight, x 10⁶ | - - - - Water Solution Properties - - - - | | | | Optimum Range of Action, pH | Maximum Work Solution Conc. Recommended, % | Maximum Conc. Recommended for Potable Water Use (by US PHS), mg/l | Manufacturer and/or Distributor (Table 4) |
| | | | | | | | Solution, % | Viscosity, cp | Density, g/ml | pH | | | | |
1	2	3	4	5	6	7	8	9	10	11	12	13	14	15
SA 1704.2													1	28
SANFLOC			Syn											95
SANPOLY K	C													65
SEDIPUR (Test) T 1	A	P	Syn	PAM	600		1	2,600 (cs)		~7.0	NEU	0.1–0.5		15
TF2	A	P	Syn	PAM	600		1	950 (cs)		~7.0	ER	0.1–0.5		15
TF	N	P	Syn	PAM	600		1	160 (cs)		~7.0	ER	0.1–0.5		15
TF5		P	Syn									0.1		15
SL6	A	L (10%)	Syn	PA										15
SL7	A	L (10%)	Syn	XP										15
TF7	A	L	Syn	XP										15
KA	C	L (30%)	Syn	PEI				1,000–5,000	1.09	7		0.1–1		15
(development product) AS	A													15
FK₃														
LK 4042														
LK 4045														
SEDOMAX F		L (13%)	Syn	PN										62
FA		L (13%)	Syn	PN										62
G		L (16%)	Syn	PN										62
(formerly DS 4101) HP	A	P	Syn	PAM										62
SEDOSAN			Syn	PAM										72
SEPARAN 2610 (= NP 10)	N	P	Syn	PN-M										28
AP30	A (N)	P	Syn	HPAM	640–801	2–3	1	3,000–4,000	1	10.5	NEU + ALK	0.5–1	1	28
AP45			Syn	PAM										28
AP273	A	P	Syn	PAM									1	28

Column numbers (as printed in the header): 1 = Flocculant Trade Name; 2–5 = Chemical Composition; 6 = Bulk Density; 7 = Molecular Weight; 8–11 = Water Solution Properties; 12–15 as labeled.

Flocculant Trade Name (1)	(2)	(3)	(4)	(5)	Bulk Density, g/l (6)	Molecular Weight, $\times 10^6$ (7)	Solution, % (8)	Viscosity, cp (9)	Density, g/ml (10)	pH (11)	Optimum Range of Action, pH (12)	Maximum Work Solution Conc. Recommended, % (13)	Maximum Conc. Rec. for Potable Water Use (by US PHS), mg/l (14)	Manufacturer and/or Distributor (Table 4) (15)
SEPARAN C-19	C		Syn	X										28
C-90	C													28
C-120	C	L	Syn	PEI		0.05–0.1	1	2.0–2.5 cks	1.054					28
NP-10	A	P	Syn	PAM	464–576	~1.0	1	175–300	1	6–7	WR	1	1	28
NP10/PWG	A	P	Syn	PAM	432–544		1	40	1	6–7		1	1	28
MG-200													1	28
MGL	N		Syn	PAM										28
PC-2			Syn	PAM										28
NP-20	N	P	Syn	PAM	368–448	2.7–4.3	1	720	1	6–7		1		28
SEPARATOR PX														104
SINK-FLOC Z-3													10	83
AZ-3													10	83
Z-4													10	83
AZ-4													10	83
AZ-5													10	83
Z-5													10	83
SOLVITOSE N	C		Nat	S										
SORBO													20	9
SP 2	C													98
3	C													98
SPEEDIFLOC 1													10	23
2													5	23
STIPIX AD			Syn	PA										37
N-80			Syn	PA										37

| Flocculant Trade Name | Chemical Composition | | | | Bulk Density, g/l | Molecular Weight, x 10^6 | Water Solution Properties | | | | Optimum Range of Action, pH | Maximum Work Solution Conc. Recommended, % | Maximum Conc. Recommended for Potable Water Use (by US PHS), mg/l | Manufacturer and/or Distributor (Table 4) |
| | | | | | | | Solution, % | Viscosity, cp | Density, g/ml | pH | | | | |
1	2	3	4	5	6	7	8	9	10	11	12	13	14	15
STIPIX K			Syn	PA										37
SUMIFLOC FC	C		Syn	PAM										
SUPERCOL (Guar Gum) GF	N	P	Nat	G	640–865		1	3,200	1	6–7		1	10	48
S 1	N	P	Nat	G	496–672		1	3,700	1	6–7		1	PWG	48
SUPERFLOC 16 (= Magnifloc)	N	P	Syn	PN-M	448–560	3–6	1	320	1	4–5		1	1	25
20	N	P	Syn	PAM	416–528		1	200–300	1	4–5		1	1	25
84	N	P	Syn	PAM	400–512		1	125–325	1	6–7		1	1	25
N-100	N	G	Syn	PAM									<0.5 (PWG)	5
N-100S	N	G	Syn	PAM										5
127	N	P	Syn	PAM	448–544		1	225–650	1	6–7		1	1	5, 25
128	N	P	Syn	PAM	473		1	750	1	6–7		1	1	25
A-100	A	G	Syn	PAM									<0.5 (PWG)	5
A-110	A	G	Syn	PAM									<0.5 (PWG)	5
A-115	A	G	Syn	PAM										5
A-125	A	G	Syn	PAM										5
A-130	A	G	Syn	PAM									<0.5 (PWG)	5
A-137	A	G	Syn	PAM										5
A-150	A	G	Syn	PAM									<0.5 (PWG)	5
C-100	C	G	Syn	PAM									<0.5 (PWG)	5
C-110	C	G	Syn	PAM									<0.5 (PWG)	5
521C	C	L	Syn	PS			Orig	225–325 (200–450)	1.15	4–5 (5.5–7.5)		<10:1 (D)	<10	5, 25
(Ser. 500—cationic) 560	C													25
570	C												10	5
571	C												10	5

Flocculant Trade Name	- - - Chemical Composition - - -				Bulk Density, g/l	Molecular Weight, x 10^6	Water Solution Properties				Optimum Range of Action, pH	Maximum Work Solution Conc. Recommended, %	Maximum Conc. Recommended for Potable Water Use (by US PHS), mg/l	Manufacturer and/or Distributor (Table 4)
							Solution, %	Viscosity, cp	Density, g/ml	pH				
1	2	3	4	5	6	7	8	9	10	11	12	13	14	15
SUPERFLOC C-573	C	L	Syn	PAM			Orig	200–500	1.14–1.18	5–7	WR	100:1 (D)	10 (PWG)	5
C-577	C	L	Syn	PAM			Orig	500–800	1.14–1.18	5–7	WR	10:1 (D)	10 (PWG)	5
C-581	C	L	Syn	PAM			Orig	4,000–6,000	1.14–1.18	4.5	WR	10:1 (D)	10 (PWG)	5
(Ser. 800—cationic) 820A	A	P	Syn	PAM	560	6	0.5	65	1	3.6 (4.2)		1		25
835A	A	P	Syn	PAM	544	15	0.5	3,300	1	7.0 (6–7)		0.1		25
836A	A	P	Syn	PAM	464	15	0.5	2,000	1	6.8 (6–7)		0.1		25
837A	A	P	Syn	PAM	480	15	0.5	620	1	5.4		0.5		25
845	A												1	5
846	A												1	5
847	A												1	5
865A(P)	A	P	Syn	PAM	800	4	0.5	20	1	3.8 (4.5)	3–10	1.0	1	25
870A	A	P	Syn	PAM	480	0.2	0.5	38	1	7.5		1.0		25
875A	A	P	Syn	PAM	448	0.4	0.5	38	1	7.5		1.0		25
880A	A	P	Syn	PAM	416		0.5	20	1	5.5		1.0		25
(Ser. 900—nonionic) 900N	N	P	Syn	PAM	528	4	1.0	150	1	4.5–5.0		1		25
901N	N	P	Syn	PAM		10	1.0	300	1	4.5–5.0		1		25
902N	N	P	Syn	PAM		7	1.0	200	1	4.5–5.0		1		25
905N	N	P	Syn	PAM		15	1.0	500	1	4.5–5.0		1		25
900P	N	P	Syn	PAM		4	0.5	25		5.0	3–10		1	25
901P	N	P	Syn	PAM		10	0.5	25		5.0	3–10		1	25
902P	N	P	Syn	PAM		7	0.5	25		5.0	3–10		1	25
905P	N	P	Syn	PAM		15	0.5	25		5.0	3–10		1	25

Flocculant Trade Name	- - - Chemical Composition - - -				Bulk Density, g/l	Molecular Weight x 10⁶	Water Solution Properties				Optimum Range of Action, pH	Maximum Work Solution Conc. Recommended, %	Maximum Conc. Recommended for Potable Water Use (by US PHS), mg/l	Manufacturer and/or Distributor (Table 4)
							Solution, %	Viscosity, cp	Density, g/ml	pH				
1	2	3	4	5	6	7	8	9	10	11	12	13	14	15
SUPERFLOC 992			Syn	PAM-CO									$<$1 (PWG)	25
SURSOLAN P5	A	P	Syn	PAM	~700						NEU + AC	0.1–1.0		15
SWIFT'S POLYMER (flocculants) X 100	C	G	Syn	PAM		4-6								102
X 110	C	G	Syn											102
X 111	C	G	Syn											102
X 112	C	G	Syn											102
X 400	A	.G	Syn	PAM		12								102
X 410	A	G	Syn											102
X 420	A	G	Syn											102
X 700	N		Syn	PAM		15					ER			102
SYNTHOFLOC 8011	N	P	Syn	PS								0.05–0.2		103
8012	N	P	Syn	PS								0.05–0.2		103
8013	N	P	Syn	PS								0.05–0.2		103
8030	A	P	Syn	PS								0.05–0.2		103
8031	A	P	Syn	PS								0.05–0.2		103
8080	AS	P	Syn	PS								0.05–0.2		103
8090	AS	P	Syn	PS								0.05–0.2		103
80 K	C	L (50%)	Syn	PS								0.05–0.2		103
STARCH (not modified)	N	P	Nat	S								1–5		49, 101
(cationic)	C	P	Nat	S										101
(anionic)	A	P	Nat	S										101
613–45	C	P	Nat	S										84
402A	N	P	Nat	S										
(oxidized)	N	P	Nat	S										

Flocculant Trade Name	- - - Chemical Composition - - -				Bulk Density, g/l	Molecular Weight × 10⁶	Water Solution Properties				Optimum Range of Action, pH	Maximum Work So-lution Conc. Recommended, %	Maximum Conc. Rec-ommended for Po-table Water Use (by US PHS), mg/l	Manufacturer and/or Distributor (Table 4)
							Solution, %	Viscosity, cp	Density, g/ml	pH				
1	2	3	4	5	6	7	8	9	10	11	12	13	14	15
TALOFLOTE			Syn	PAM										
TOLFLOC 300		L												90
340	A		Syn								WR			90
TR-Flocculant ZAP	A	P	Syn	HPAM-S	550–750									106
11APR	A	P	Syn	HPAM	550–750									106
(= low MW 95AP) 25AP	A	P	Syn	HPAM	500–600						4–12	0.1		106
90AP	A	P	Syn	HPAM	550–750	$\geqslant$10					(opt. 8)	0.1		106
91AP	A	P	Syn	HPAM	850–900	$\geqslant$10						0.1		106
93AP	A	P	Syn	HPAM-S	550–750	$\geqslant$10						0.1		106
95AP	A	P	Syn	HPAM-S	550–750	$\geqslant$10						0.1		106
80NP	N	P	Syn	PAM	500–600	10					ER	0.1		106
80NPZ	N	P	Syn	PAM	500–600	10					ER	0.1		106
90CP	C	P	Syn	(PAM)	500–650	10+					3–8(6)	0.1		106
91CP	C	P	Syn	(PAM)	500–650	10+					3–8(6)	0.1		106
11CPR	C	P	Syn	(PAM)	550–750						3–8(6)			106
CLR	C	L (20%)	Syn	PAN-A					$\sim$1					106
(sample) XCP	C	P	Syn											106
(sample) YCP	C	P	Syn											106
TRAGAFLOK	N	P	Nat	S-P							10	2–5		3
TRETOLITE (cca 20 types)		L	Syn											90
TYDEX L 12														
14		L					$\sim$1			7.8				
1L	C		Syn	PAM										
TYCHEM 8013	N	P	Syn	PAM	576		1	750	1	6		1		99

Flocculant Trade Name	- - - Chemical Composition - - -				Bulk Density, g/l	Molecular Weight × 10^6	Water Solution Properties				Optimum Range of Action, pH	Maximum Work Solution Conc. Recommended, %	Maximum Conc. Recommended for Potable Water Use (by US PHS), mg/l	Manufacturer and/or Distributor (Table 4)
							Solution, %	Viscosity, cp	Density, g/ml	pH				
1	2	3	4	5	6	7	8	9	10	11	12	13	14	15
TYCHEM 8021	C													99
8024	N	P	Syn	PAM	640		1	200	1	5.5–6		1	1	99
8030	A													99
8034													1	99
8035													1	99
TYLOSE CBR 4000		P	Nat	CP										
U.C. 19														108
UCAR RESIN C-149	C	P	Syn	PAN-HS	288–368		1	20–100	1	3–4		1		108
UETIKON-flocculants A 3	A	P	Syn	PS	690–800							≥4.0		58
K 3	C	L (30%)	Syn	PS					1.12			≤12		58
K 4	C	P	Syn	PS	610–710						WR			58
N 3	N	P	Syn	PS	590–730						ER			58
URONOFLOK	A-N		Nat											19
VA 2	C	L	Syn	PVM								D		98
3	C	L	Syn	PVP								D		98
(4, 6, 7, 10, 11, 12, 13, 15)	C		Syn											98
2K														98
2M														98
2T, 3T	C		Syn	PVT										98
102	C		Syn	PMA										98
212	C		Syn	PMA										98
VAPORIN			Syn	PAN									<7	111
VAPTREAT H			Syn	PA									<10	3a, 56

Flocculant Trade Name	Chemical Composition				Bulk Density, g/l	Molecular Weight × 10⁶	Water Solution Properties				Optimum Range of Action, pH	Maximum Work Solution Conc. Recommended, %	Maximum Conc. Recommended for Potable Water Use (by US PHS), mg/l	Manufacturer and/or Distributor (Table 4)
							Solution, %	Viscosity, cp	Density, g/ml	pH				
1	2	3	4	5	6	7	8	9	10	11	12	13	14	15
VARCO-FLOC													150	109
VERDICKUNG AN		L	Syn	PA					1.1	6.0		0.01		110
WELGUM S	A	P	Nat	AL								0.05	<1	1
WISPROFLOC (= D 100)														
20	N	P	Nat	S							WR	1–2	<5	96, 112
A	A	P	Nat	S								1–2		96, 112
P	C	P	Nat	S								1–2	<3	96, 112
XD (name change) 7817												1	1	28
ZETA-FLOC C													20	83
K													20	83
O													20	83
WA													20	83
ZETAG 32	C	P	Syn	PAM										3
51	C	L (20%)	Syn	PAN										3
63	C	P	Syn	PAM										3
92	C	P	Syn	PAM										3
94	C	L (15%)	Syn	PAN										3
ZIMMIE 102			Syn								WR			113
ZIMMITE 120	A	P	Syn	PAM			Orig		1.1			D		113
ZM-100													1	113
ZC-301													130	113
ZT-600											WR		1	113
ZT-601													1	113
ZT-603													1	113

Flocculant Trade Name	- - - Chemical Composition - - -				Bulk Density, g/l	Molecular Weight x 10^6	- - - - Water Solution Properties - - - -				Optimum Range of Action, pH	Maximum Work Solution Conc. Recommended, %	Maximum Conc. Recommended for Potable Water Use (by US PHS), mg/l	Manufacturer and/or Distributor (Table 4)
							Solution, %	Viscosity, cp	Density, g/ml	pH				
1	2	3	4	5	6	7	8	9	10	11	12	13	14	15
ZT 600–604			Syn	PAM	480								PWG	113
650	A	P	Syn	XP	~980				1	4		0.5–1.0		113
ZUCLAR 110 PW													0.5	35

Explanatory Notes

Column 1

Polyflok 37cP diluted by water in the ratio of 1:1 by weight gives Polyflok 3750P

Column 2

A—anionic
C—cationic
N—nonionic
AC—amphoteric
L—low viscosity
M—medium viscosity
ML—viscous type (Reten)

Only basic predominant ionic characteristics are given. Individual products have different degrees of hydrolysis and are either strongly, intermediately, or weakly anionic and/or cationic. Since the authors do not know the degree of hydrolysis for all products, a detailed classification cannot be given. In many cases the number of the product indicates the degree of hydrolysis, for example:

BTI—anionic character increases A 100 → A 150
Decabloc—anionic character increases A 1 → A 5
Superfloc—anionic character increases A 100 → A 150
TR-flocculants—anionic character increases 90 AP → 95 AP
Decabloc—cationic character increases C 1 → C 2
Superfloc—cationic character increases C 100 → C 110

Column 3

B—blocks
G—granules
L—liquid, usually viscous
P—powder or flakes
Pa—paste

Column 4

Nat—based on natural product
Syn—synthetic product

Column 5

AG—alkyl-guanidineamine complex
AL—derivatives of alginic acid Na polyalginate
AP—mixture of basic aluminum chloride and high molecular weight synthetic polymer
APC—alumina, polymer and caustic soda
APD—aminopolysaccharide from sea crustacea scrap
BIC—biocolloid, inorganic coagulant and caustic soda
CB—extract from *Cystoseira barbata*
CH—product of seaweed
CP—carboxymethylcellulose (sodium salt)
CPN—chemically modified natural polymer
DAMA—polydialkyldimethylammonium chloride
DAMA-M—modified polyalkyldimethylammonium chloride
G—derivatives of guar gum
HEC—hydroxyethylcellulose
HPAM—polyacrylamide (partially hydrolyzed)
HPAM-S—polyacrylamide (highly hydrolyzed)
KH—carbohydrate
KO—isinglass fines
KODT—condensation product of hexamethylenediamine, dichloroethane and olefinic acid

Column 5 (continued)

L—leguminous seed derivates, natural polymer

MA—copolymer of methacrylic acid and maleic anhydride

Mi—extract of mimosa (tannin-based)

MVM—copolymer of methyl vinyl alcohol and maleic anhydride

PA—polyacrylate of sodium or ammonium (derivatives of polyacrylic acid)

PA-A—polymer of acrylic acid containing ionic (amine) and nonionic (amide) groups in various ratios

PA-AA—flocculants based on polyamines, polyamides, polyacrylamides

PA-B—polymer of acrylic acid with reactive polyamide groups

PA-BC—polymer of acrylic acid containing varying portions of ionic (carboxylic-COONa) and nonionic (amide-NH_2) groups

PAC—polyaminocarbonic acid ester

PAD—polyamide

PAM—polyacrylamide (or polyacrylamide-based)

PAM-AN—mixture of polyacrylamide-based polymer and inorganic salt

PAM-C—copolymers of polyacrylamide

PAM-CO—acrylamide and acrylic-acid-based copolymers

PAM-COA—copolymers of polyacrylic acid, acrylamide and acrylate

PAM-CON—copolymers of sodium acrylate, acrylamide and acrylonitrile

PAM-S—substituted polyacrylamide

PAN—polyamine (usually quaternary ammonium group) or flocculant of polyamine type

PAN-A—polyamine-amide-based

PAN-H—condensation product of alkylene-polyamine and polyfunctional halohydrin

PAN-HS—condensation product of alkylene-polyamine and polyfunctional halohydrin (product contains sulfo groups)

PEI—polyethyleneimine

PEN—polyethylene oxide polymer

PMA—polymethacrylates-sodium salt (and/or hydrolyzed)

PN—polyacrylonitrile

PN-H—polyacrylonitrile (hydrolyzed)

PN-M—modified polyacrylonitrile

PS—synthetic high molecular weight polymer

PS-N—synthetic high molecular weight polymer with quaternary ammonium group

PS-T—polystyrene

PVM—poly(4-vinyl-N-benzyltrimethylammonium) chloride

PVP—poly(4-vinyl-N-benzylpyridinium) chloride

PVT—polyvinyltoluene

S—starch-based derivatives

S-C—mixture of starch derivative with calcium chloride

S-CM—carboxymethyl starch

S-P—pregelled starch

S-S—mixture of starch derivatives

SM—organic polymer (XP) and bentonite

SM-AN—organic polymer (XP) and aluminum hydroxide

SP—sulfonated polymer

T—tannin derivatives

TC—synthetic polymer and caustic soda

VAM—copolymer of vinyl acetate and maleic anhydride

X—products without any specification

XP—high molecular weight organic polymer (without further specification)

XP-AN—high molecular weight organic polymer (without further specification) and inorganic salt

XPS-C—mixture of high molecular weight cationic polymers

Column 6

Bulk density varies according to the sample, that is, whether loose, packed or in the working state

Column 8

Orig—product as delivered

Column 9

Viscosity varies with time, pH, temperature, rpm, etc.

Column 10

In most cases these are approximate values

Column 11

pH values given are in distilled water; solutions in supply water generally show a little higher pH

Column 12

AC—effective in acid range of pH
ALK—effective in alkali range of pH
ER—effective within the entire range of pH
NEU—effective in neutral range of pH
WR—effective over a wide range of pH
The pH value indicated in parentheses is the optimum pH value for flocculant efficiency

Column 13

With liquid flocculants sometimes only a dilution ratio is given (D)

Column 14

Flocculants approved by US PHS for potable water use (U.S. PHS Technical Advisory Committee on Coagulant Aids for Water)

Column 15

See Table 4. Manufacturers as well as distributors are given in some cases. European representatives of American manufacturers are listed but it must be understood that these commercial relationships are often quite transitory.

Table 3

Use of Organic Flocculants
in Water Treatment

Organic Flocculant Trade Name	Public and Industrial Water Treatment	Application — Municipal (A) and Industrial (B) Wastewaters	Sludge Treatment and Industrial Suspensions
ACCOFLOC 305		(B) Flocculation of wastewaters from rice washing, wastewaters from filter washing after sugar juice filtration	Paper and cellulose production
AEROFLOC 3453			Selective flocculation in ore processing industry
AQUAFLOC 408	Coagulant aid in clarification of raw water for potable use ($<$50 ppm), and softening processes	(A)(B) Coagulant ($\sim$25–300 ppm), and coagulant aid (5–100 ppm) in industrial and municipal waste treatment	Thickening of sludges
409	Coagulant aid in clarification of raw water for potable use, and softening processes	(B) Separation in industrial waste treatment (0.1–2.0 ppm)	Thickening of sludges
410	Flocculant in softening processes	(B) Primary coagulant—replaces the use of ferric and alum salts (5–50 ppm); May be used in conjunction with 10–30 ppm of bentonite clay; Coagulant aid (0.5–10 ppm); Emulsion breaker (10–200 ppm)	Sludge conditioner prior to sludge dewatering in centrifuges and vacuum filters—dosage 0.25–4% of dry solids
412		(B) In a wide variety of liquid-solid separations (20–100 ppm)	
414		(B) In a wide variety of liquid-solid separations, esp. in flocculation of iron-bearing wastewaters (0.5–5.0 ppm)	
418		(A)(B) Primary coagulant (1–50 ppm), coagulant aid (0.5–10 ppm)	Dewatering of all types of municipal and industrial sludges (centrifuges, vacuum filters—dosages 0.24–0.03 kg/m^3 of sludge treated
419		(A)(B) Primary coagulant (1–50 ppm); Coagulant aid (0.5–10 ppm); Esp. recommended for use in wet scrubber systems in foundries	Sludge conditioning and dewatering from municipal and industrial waste treatment (thickeners, centrifuges, vacuum filters—dosage 0.24–0.03 mg/m^3 of sludge treated

Organic Flocculant Trade Name	Public and Industrial Water Treatment	Municipal (A) and Industrial (B) Wastewaters	Sludge Treatment and Industrial Suspensions
		--Application--	
AQUAFLOC 421		(A)(B) Coagulant aid for a wide variety of wastewater treatment operations (0.1–5 ppm); By phosphate removal in conjunction with 50–100 ppm of iron or aluminum salts followed by 0.5–2 ppm of flocculant	
431		(B) Oil-water emulsion breaker	Sludge conditioning
ARAFLOC A-185		(B) Flocculation of wastewaters from paper and cellulose production	
ARFLOC			Sludge conditioning
BETZ POLY-FLOC ser. 1100	Plant influent water clarification; In-line clarification; Industrial process water clarification	(A)(B) Sewage and industrial waste treatment (settling, air flotation, centrifugation)	Sludge thickening and dewatering by vacuum filtration, on sand beds; Liquid/solids separations in the processing of magnesium oxide, titanium dioxide, lime, aluminum sulfate, phosphoric acid, and several other chemicals
1110		(B) Improvement of primary sedimentation of refinery wastewater (consisting of API separator, effluent combined with miscellaneous plant wastes and sanitary sewage, dosage $\sim$ 1 ppm)	
1130		(B) Clarification (sedimentation) of steel mill–rolling mill and pickle acid wastes ($\sim$0.5 ppm)	
1160	Fresh water clarification for paper-making processes, reduced dosage of alum ($\sim$0.3 ppm)		
1170			Titanium dioxide processing: conditioning of acid leach liquor for filtration
1200	Clarification of turbid surface waters in conjunction with inorganic coagulants;	(A)(B) Improvement of sedimentation efficiency of industrial and municipal primary waste (0.2–5.0 ppm)	

Organic Flocculant Trade Name	Application		
	Public and Industrial Water Treatment	Municipal (A) and Industrial (B) Wastewaters	Sludge Treatment and Industrial Suspensions
BETZ POLY-FLOC ser. 1200		(B) Settling of iron oxide suspensions from steel industry wastewaters (0.2–5.0 ppm)	
BTI (generally)	Potable water clarification: Initial clarification and sedimentation (as adjuncts to the traditional coagulants, dosage: 0.01–0.1 ppm); Treatment of filter feed (gravity and pressure filters, in conjunction with alum, dosage: 0.01–0.1 ppm); Clarification of back washings from filters; Treatment to condition water for processes systems; Pretreatment of boiler feedwater and cooling water (alone or in conjunction with primary coagulants)	(A) Digester supernatant clarification; (B) Clarification of industrial waste, e.g., iron and steel works, steelworks blast furnace gas washing, steelworks pickling lime, food factory, tannery effluents (non-sulfide, sulfide)	Sludge treatment from sedimentation/ clarification processes (potable water, dosage 1–20 ppm); Sludge (sewage treatment) thickening by gravity, flotation and centrifuging); vacuum filtration of sludges, sludge dewatering on drying beds, secondary sludge settling, clarification of activated sludge prior to river discharge
A100	Coagulant aid in clarification with Al salts		Improvement of sludge filtering properties from up-flow units, improvement of filterability and dewatering of sludge from tannery wastes treatment plants (sulfide effluents), flocculants in copper ore processing plants (for thickening of flotation tailings); Paper industry: retention aid
A110	Coagulant aid		Sludge dewatering (drying beds), flocculant in Cu-ore processing plants (for thickening tailings prior to acid leach of solubles); Sugar industry: sugar juice clarification, filter aid for sugar muds; Paper industry: retention aid
A130	Coagulant aid		Sugar industry: sugar juice clarification, filter aid for sugar muds; Flocculant in iron-ore mining (at any stage of washing/concentrating process

Organic Flocculant Trade Name	Application		
	Public and Industrial Water Treatment	Municipal (A) and Industrial (B) Wastewaters	Sludge Treatment and Industrial Suspensions
BTI A150	Coagulant aid in conjunction with primary coagulant	(A) Improved suspension sedimentation in raw sewage (0.5–1.0 ppm)	Sewage treatments: primary sludge settling; Sugar industry: clarification of sugar juice and filter aid for sugar mud; Improvement of filterability and dewatering characteristics of sludges, e.g., from clarification tannery wastes (sulfite, nonsulfite effluents)
A12	Coagulant aid		Red mud flocculant in Bayer aluminum plants (mud separation and mud washing operations)
N100	Coagulant aid		Flocculant in many copper ore treatment plants (chalcopyrite, pyrites)—sedimentation/thickening operations, thickening of concentrates
C100			Flocculant in copper ore treatment plant (e.g., chrysocolla) (80 g/ton of dry solids);
C110			Paper industry: retention aid
CL 40	Removal of color forming substances from water; In-line clarification (5–10 ppm)		Paper industry: retention aid Settlement of accelerated sludge (5–10 ppm)
CR 50 LP 150		(A) Sewage purification	Paper industry: retention aid
CALGON Coagulant Aid (generally)	Clarification of potable and nonpotable water supplies; Cold and hot process softening; Industrial water systems	(A)(B) Clarification of municipal and industrial wastewaters	Conditioning of sewage sludges; Mineral processing application (wastes, tailings); Recovery of coal fines after washing operation; Brine clarification; White water separation; Ore beneficiation (thickening concentrates and tailings in extracting metals from their ores); High solids-liquid separations

Organic Flocculant Trade Name	Public and Industrial Water Treatment	Municipal (A) and Industrial (B) Wastewaters	Sludge Treatment and Industrial Suspensions
		----- Application -----	
CALGON Coagulant Aid 2 11	Clarification of potable water supplies (max. 1 ppm)		Clarification in the secondary treatment up-flow unit by recovery of bromine from oil field brines
18	Clarification of potable water supplies (max. 15 ppm); hot process lime-soda softening (8–12 ppm); combined treatment with No. 18 (before or with coagulant 2–3 ppm) and No. 952 (after coagulant 0.25–0.5 ppm) for potable applications; for nonpotable applications: combination of No. 18 (before or with coagulant, 2–3 ppm) and No. 50 (after the coagulant 0.15–0.25 ppm); combination of No. 223 (before the coagulant, 1–3 ppm) and No. 18 (with or immediately after the coagulant, 5–20 ppm); combination of No. 223 as primary coagulant (2–3 ppm) added first, followed by No. 18 (15–20 ppm)	(B) Combination of No. 18 (dosage 10–20 ppm) added first, followed No. 50 (0.5–2.0 ppm) for metallic wastes; combination: No. 223 (2–3 ppm) added first, followed No. 18 (15–20 ppm) for wastes containing oil	For high solids-liquid separation (5–10 ppm)
37		(B) Treatment of blast furnace gas washing wastes	
225	Primary coagulant or coagulant aid in clarification of industrial water supplies (0.5–3.0 ppm); Cold lime-soda softening processes	(A) Coagulant aid in primary settling tanks and final effluent clarifiers (0.25–3.0 ppm) or as primary coagulant (1–5 ppm)	Sludge conditioning (raw or activated sludge) for dewatering processes (drying beds), mechanical dewatering equipment (alone or in conjunction with anionic or nonionic sludge conditioners 3–10 kg/ton of dry solids); elutriation and air flotation processes (raw and activated sludge) (50–100 ppm); Improving clarification in high or low solids/liquid preparation processes
226	Primary coagulant or coagulant aid in industrial water clarification (0.5–2.0 ppm);	(A) Coagulant aid in primary settling tanks and final effluent clarifiers (0.5–1.5 ppm) (with or without	Conditioning of sewage sludges for dewatering processes (drying beds, centrifuges, mechanical dewatering equip-

Organic Flocculant Trade Name	Public and Industrial Water Treatment	Municipal (A) and Industrial (B) Wastewaters	Sludge Treatment and Industrial Suspensions
		--------Application--------	
CALGON Coagulant Aid 226	Cold lime-soda softening processes	primary coagulants)	ment, alone or in conjunction with anionic or nonionic sludge conditioners, 1–4 kg/ton of dry solids); elutriation or air flotation of sewage sludges (20–60 ppm)
230	Coagulation aid for clarification of nonpotable water (0.1–0.5 ppm)		For high solids-liquid separation processes (4.5–90 g/ton of solids)
240	Coagulant aid in industrial water clarification (0.1–0.5 ppm)		For high solids-liquid separation processes (4.5–90 g/ton of solids); Coal washery waters
233, 243, 253	Coagulant aid in industrial water clarification (0.1–2.0 ppm); Lime-soda softening operations (particularly effective: No. 253)	(A) Coagulant aids in primary settling tanks and final effluent clarifiers, with or without primary coagulants (0.25–1.5 ppm)	Conditioning of sewage sludges for dewatering (drying beds, mechanical dewatering equipment, dosage 0.5–1.0 kg/ton of dry solids; alone or in conjunction with cationic sludge conditioners); conditioning of sewage sludge ahead of centrifuging (0.5–1.5 kg/ton of dry solids); elutriation or air flotation (5–50 ppm)
2250	Primary coagulant (1–5 ppm) or a flocculant (0.25–3.0 ppm) in the clarification of raw water supplies (may be used either alone or in conjunction with other primary coagulants, or with special clays as Calgon Coagulant Aid No. 25); Cold lime-soda softening processes		
2256, 2260 2270	Primary coagulants and flocculants in the clarification of raw river water supplies (may be used either alone or in conjunction with inorganic coagulants or clays); Cold lime softening plants		
2325	Coagulant aid in the clarification of raw water supplies (in conjunction with primary coagulant—inorganic or cationic organic—removal of turbidity or color); Lime softening		

Organic Flocculant Trade Name	Public and Industrial Water Treatment	Application — Municipal (A) and Industrial (B) Wastewaters	Sludge Treatment and Industrial Suspensions
CALGON Coagulant Aid 2350, 2400, 2425	Coagulant aids in the clarification of raw water supplies (in conjunction with an inorganic or cationic organic coagulant) (0.1–0.5 ppm); Coagulant in lime or lime-soda softening processes, particularly in hot process industrial softeners (No. 2350, widespread use)		
CARAFLOK 91AP	Coagulant aid (sand and clay suspensions);		
95AP	Coagulant aid (sand and clay suspensions)		
CAT-FLOC	Primary coagulant and/or coagulant aid in clarification of potable or industrial water supplies; (sedimentation, sludge blanket stabilization; dosage as primary coagulant 0.2–7.0 ppm; as coagulant aid 0.10–2.0 ppm); higher dosage may be required in case the water has a high pH value or contains anionic detergents, lignins, phosphates or high total solids	Clarification of wastewaters	
B	Primary coagulant in the clarification of municipal and industrial water (sometimes with specially selected clay, sometimes with inorganic coagulants—improved clarity, color removal, reduction of inorganic coagulants by 40–60%)		
DECABLOC (generally)	Treatment of raw water	(B) For sand and gravel washing, coal tailings disposal, tannery effluents, acid pickling wastes and plating wastes	In mining operations; for sludge dry- and disposal

Organic Flocculant Trade Name	------------------------------Application------------------------------		
	Public and Industrial Water Treatment	Municipal (A) and Industrial (B) Wastewaters	Sludge Treatment and Industrial Suspensions
DECAPOL A 30 P			Solids separation of $CaCl_2$–NaCl waste solutions from soda production
DELFLOC (see HERCOFLOC)			
DOW A-22		(B) Cellulose wastewater purification, tannery wastewater containing liming wastes	
A-23			Sludge conditioning
DREWFLOC 1	Coagulant aid in conjunction with inorganic salts in potable water treatment, color and turbidity removal; Settleability improvement in cold and hot softening processes (2–10 ppm)		
3	Coagulant aid in raw water clarification (0.25–5.0 ppm)	Waste disposal application Municipal water clarification	Liquid-solids separation in many industrial processes (pigment production, liquid detergents production, mineral beneficiation, etc.)
5	Coagulant aid in raw water clarification (0.1–2.0 ppm) Primary coagulant in specific clarification applications (comparatively high content of dissolved solids, dosage 0.5–5.0 ppm)		
21	Coagulant or coagulant aid in raw water clarification (0.25–5.0 ppm)	(A) Coagulant or coagulant aid in industrial wastewater treatment	Sugar industry: raw cane juice clarification
31	Primary coagulant or coagulant aid in raw water clarification (0.01–0.5 ppm)	Waste disposal	Solids-liquid separation in industrial processes
35	Primary coagulant (2–15 ppm) or coagulant aid in water clarification (0.3–2.0 ppm)	Waste disposal	Industrial processes
138		(A) Flocculation of sewage in domestic sewage plants	Dewatering of sludge in domestic sewage plants (replacement of ferric chloride, lime; vacuum filtration, centrifug-

Organic Flocculant Trade Name	Public and Industrial Water Treatment	Application Municipal (A) and Industrial (B) Wastewaters	Sludge Treatment and Industrial Suspensions
DREWFLOC 138			ing, 0.25–1.0% of dry solids); coagulant in digested and activated sludge plants with all types of filtration equipment
139		(A) See No. 138	See No. 138
ser. 400	Primary coagulant in water clarification	Waste disposal	Liquid-solids separation in industrial manufacturing processes; Sludge conditioning before incineration; Sludge thickening (vacuum filtration, centrifuges, sludge drying beds, dosage level 0.5–2% based on dry solids)
DT 120 220		(A) Municipal sewage treatment; (B) Purification of various industrial wastewaters (e.g., metal-working industry)	Coal and ore processing; Retention aid in paper industry; Filter aid for wastewater sludges
FILTAFLOK	Coagulant aid (0.05–1.0 ppm); Filtration aid (0.01–0.1 ppm)	(B) Improvement of primary sedimentation (2–10 ppm)	Sludge consolidation in sedimentation tanks for water treatment process (1–5 g), filter backwash water treatment (1–5 g), hydroxide sludges dewatering on drying beds (1–5 g/l of sludge), vacuum filters (1–5 g/l of sludge), filter presses (2–5 g/l of sludge); Dewatering of sludge (raw and digested) from municipal sewage treatment plants–drying beds (1.5–3.0 g/l of sludge), vacuum filter (1.5–3.0 mg/l of sludge), filter press (1.5–3.0 mg/l of sludge), centrifuges (1.5–4.0 mg/l of sludge); Sludge from water softening–dewatering on centrifuges; Industrial suspensions thickening: through-flow thickeners (0.01–0.1 kg/ton of dry solids), vacuum filters (0.01–0.1 kg/ton of dry solids, e.g., fine coal), filter press (0.1–0.2 kg/ton of dry solids), centrifuges (0.2–0.5 kg/ton of dry solids), sedimentation of tailings in lagoons (0.01–0.05 kg/ton of dry solids)

Organic Flocculant Trade Name	Application		
	Public and Industrial Water Treatment	Municipal (A) and Industrial (B) Wastewaters	Sludge Treatment and Industrial Suspensions
FILTAFLOK A1			As an aid to the filtration of fine coal on rotary vacuum filters
FLOC-ACE			Flocculation of cellulose slurry from paper production
FLOCCOTAN CS7			Conditioning of primary digested sludge before mechanical dewatering
FLOCONIT	Flocculation of suspensions in water, clarifying process acceleration		
GAMLOSE		(B) Coal washery effluents, deoiling of wastewaters; flocculation of liquids from manufacturing processes, ore processing (e.g., U ore)	Sludge thickening
GIGAMID			Flocculation of Zn slurries
GIGTAR			Flocculation of Cu concentrates
GIPAN			Flocculation of coal slurries
HERCOFLOC (generally)	Treatment of water for potable and industrial use	(A) Flocculation of municipal wastewater; (B) Purification of industrial wastewaters from iron and steel processing, automobile industry, cement production, stone coal, rock oil, chemicals, ceramics, glass, phosphorus, rubber, textile, leather, food industries, plating wastewaters	Sludge dewatering (primary raw sludge 0.5-1 kg/dry solids, other sludges 2 kg per ton of dry solids); vacuum filter, centrifuges, filter press, sludge drying beds, screen centrifuges; Solids-liquid separation in industrial processes (e.g., sugar, pigment, mineral industries)
800, 810, 814	As flocculants		Industrial suspension separation; Thickener and conditioning agent for sludge dewatering (vacuum filters, filter press, centrifuges, drying beds)
816, 819, 824, 831			Flocculant and thickener in a variety of industrial applications; conditioning

Organic Flocculant Trade Name	--Application--		
	Public and Industrial Water Treatment	Municipal (A) and Industrial (B) Wastewaters	Sludge Treatment and Industrial Suspensions
HERCOFLOC 816, 819, 824, 831			agent for sludge dewatering (in conjunction with cationic polymers)
HYPAN			Clarification of coal slurries, coal flotation wastewater
JAGUAR (and/or Reagent MRL) MDD, WPD, J2S1, MRL-22, 705 ser., 800 ser.	Water treatment for nonpotable use (WPD 0.10–0.50 ppm)	Industrial wastewater purification	Clarification of industrial suspensions up to 2 g/l (0.50–5.0 ppm); thickening aid in working of ores, metallurgy; Sedimentation of mine water (7–226 g per ton of dry solid), filtration thickening (22–226 g/ton of dry solids)
MDD			Filter aid for acid and alkaline leached uranium ores, acid leached copper ores, cyanide gold ores, etc. (22.5–225 g/ton of dry solids); Settling and clarifying aid for acid leached uranium and copper ores, magnesium hydroxide, iron ore, clay, etc. (0.50–5.0 ppm); Thickening (~4.5–225 g/ton of dry solids); Flotating (~4.5–225 g/ton of dry solids)
500 ser. MDB, WPB	See the first group of JAGUAR		
MD-7A	See the first group of JAGUAR		
MRL-22-A	See the first group of JAGUAR	Clarification of white water (paper)	Clarification of dilute slurries (0.025–5.0 ppm); Thickening of (~4–22 g/ton of dry solids); Flocculation of kaolin type clays, ore slimes
MRL-19, MRL-159	See the first group of JAGUAR		
MRL-295			Thickening of industrial suspensions (2–113 g/ton of dry solids), flocculation of diluted suspensions (0.025–2.5 ppm), filtration (4–113 g/ton of dry solids) (ore processing and metallurgy)

Organic Flocculant Trade Name	Application		
	Public and Industrial Water Treatment	Municipal (A) and Industrial (B) Wastewaters	Sludge Treatment and Industrial Suspensions
JAGUAR MRL-13, MRL-14, MRL-93, MRL-364, MRL-95	Nonpotable water treatment (MRL-13, MRL-14, MRL-93; 0.10–1.0 ppm)		Thickening (20–450 g/ton of dry solids), filtration (90–450 g/l), flocculation of industrial suspensions (0.50–10 ppm) (ore processing and metallurgy)
MRL-91		Clarification aid (in conjunction with Stein, Hall Polyhall products), e.g., wastewaters from Fe ore processing	Thickening of iron ore tailings (4–22 g per ton of dry solids); Flocculation of diluted suspensions (0.5–5.0 ppm); Filtration of industrial suspensions (22–226 g/ton of dry solids), flotation (22–226 g/ton of dry solids)
K-4			Flocculation of corundum water suspension, flocculation of precipitated heavy metal hydroxides
KOMETA			Flocculation of coal slurries
KURIFLOC			Papermaking industry
MAFLOC			Municipal sewage sludge dewatering (drying beds, vacuum filters, filter press)
MAGNAFLOC (generally)	Potable water treatment, cooling water treatment, clarification of underground water, clarification of process water, industrial water treatment, clarification of township water	(A) Clarification of sewage; (B) Clarification of cooling water and scrubber effluents in the iron and steel industry, electroplating effluents, furnace gas washing, textile effluents, paper and pulp effluents, coal washing, clarification of effluents and wash waters in sand and gravel industry	Dewatering of sludges (filter pressing, vacuum filtration, lagooning, drying or under-drained beds); Sedimentation/thickening and clarification of suspensions; Industrial applications: mineral and mining processing (ferrous, nonferrous and nonmetallic minerals—dewatering of concentrates, thickening of tailings), density adjustment and desliming of process pulp; Hydrometallurgy, various solid-liquid separation stages (preparation of pulp prior to leaching—acid, alkaline and aqueous leach, cyanidation, separation of pregnant liquor from leach residue,

Organic Flocculant Trade Name	Application		
	Public and Industrial Water Treatment	Municipal (A) and Industrial (B) Wastewaters	Sludge Treatment and Industrial Suspensions
MAGNAFLOC			separation of pregnant liquor from purification precipitants, separation of precipitated value from barren liquor, deposition of metal in electrowinning, recovery of residual values from barren liquor); Sugar juice clarification; Flocculants in titanium dioxide manufacture; Flocculants in the manufacture of phosphoric acid and phosphates; The use in kraft mill recausticization plants (paper and pulp effluents); Retention on screens (paper, pulp and board manufacture); In industries: aluminum sulfate, cement—wet process, ceramic and clay, common salt, drugs and antibiotics, photographic, pigment manufacture, potash, seawater magnesia
MAGNAFLOC LT, Type LT series—application in water treatment plants	Coagulant aid in Al and Fe salts clarification	Flocculation of sand filter backwashes	Dewatering of waterworks sludge from both sedimentation systems and filter backwashes (drying beds and infiltration processes)
LT 25			Filtration in sugar industry
R14, Type R series—application in ore processing industry			For precipitation under acidic conditions, or of solids of an organic nature
R135			Mining and ore processing industry: thickening and reclamation of water from Fe sediments (20 g/ton)
MAGNAFLOC R 139			Dewatering in solid bowl centrifuges and hydrocyclones; Sedimentation processes (effective for organic solids);

Organic Flocculant Trade Name	— — — — — — — — — — — — — — — — — Application — — — — — — — — — — — — — — — — —		
	Public and Industrial Water Treatment	Municipal (A) and Industrial (B) Wastewaters	Sludge Treatment and Industrial Suspensions
MAGNAFLOC R 139			Hydrometallurgy: leach residue—recovery and clarification of pregnant liquor, e.g., nickel—alkaline leach (45 g/ton of dry solids)
R 140			Dewatering of thickened Cu ore concentrates (solid bowl centrifuges and hydrocyclones, 9 g/ton of dry solids); Copper electrorefining (0.2 ppm); Sedimentation process, use on organic solids or inorganic solids with adsorbed organics, particularly useful under strongly acidic conditions (acid leaching process—0.09 g/ton of pulp)
R 155	Treatment of raw water supplies	(A) Improvement of primary sedimentation	Conditioning of sludge with inorganic coagulants content or with lime content; Sedimentation and filtration processes (high solids concentration) by the production of magnesium, nickel, aluminum hydroxides, calcium, zinc, and similar carbonates; Dewatering by solid bowl centrifuges and hydrocyclones; Metal recovery by precipitation (e.g., Co acid leach, Cl acid leach, 4.5 g/ton); Barren liquor residual value recovery: manganese from uranium liquor; Leach residue—recovery and clarification of pregnant liquor: electrolytic Zn (also in the lead, cadmium, and other metal by-products recovery sections); Density adjustment and desliming of process pulps: (Cu ores, 9 g/ton of dry solids); Thickening and dewatering of Pb-Zn ore concentrates (increase by 22%, 10 g/ton of dry solids); Thickening and filtration of sludges from Cu mining wastewaters (20 g/ton)

Organic Flocculant Trade Name	Public and Industrial Water Treatment	-------------------Application------------------- Municipal (A) and Industrial (B) Wastewaters	Sludge Treatment and Industrial Suspensions
MAGNAFLOC R 156		(A) Flocculation and phosphate removal (in conjunction with lime and Fe or Al salts)	Sedimentation and filtration processes, involving mineral or other inorganic suspensions (by the production of magnesia, nickel, aluminum and other hydroxides, calcium, zinc and similar carbonates); Thickening of flotation tailings (0.022 g/ton of coal fines); Filtration of flotation concentrates; Clarification of $CaCO_3$ solutions
R 292			Sludge thickening (filter presses, rotary vacuum filters, coil filters)
R 351			Preparation of pulp, e.g., gold-cyanidation (clarify and improve recovery of pregnant liquor following cyanidation); Purification circuit–clarification of pregnant liquor, e.g., copper acid leach ($\sim$20 g/ton of suspension); Density adjustment and desliming of K_2CO_3 solution (20 g/ton of sludge)
R 352			Leach residue—recovery and clarification of pregnant liquor, e.g., uranium acid leach
R 455			Leach residue—recovery and clarification of pregnant liquor, e.g., copper aqueous leach ($\sim$4 g/m^3)
E 20		(B) Treatment of wastewaters from mining and mineral processing plants (5 g/ton)	Thickening of settled residue by the treatment of wastewater from Pb, Zn production ($\sim$5 g/ton of suspension)
EM 15			Flocculation of pulps and slips of high solids concentration
EM 127			Undiluted and unneutralized pulps of high solid concentration (simple sedimentation process, improving vacuum or pressure filtration characteristics, e.g., ball milled ceramic slips dewatered by filter pressing ($\sim$450 g emulsion per ton of solids)

Organic Flocculant Trade Name	Public and Industrial Water Treatment	Municipal (A) and Industrial (B) Wastewaters	Sludge Treatment and Industrial Suspensions
		-------- Applications --------	
MAGNAFLOC HR 120			For use on organic and mineral solids and at high acidity flocculation of bacterial suspensions and mycelial fermentations
METAS			Flocculation of coal slurries
NALCO (generally)	Potable water softening and clarification; Metal processing water; Paper mill supply water clarification	(A) Sewage treatment—primary settling (digesters, trickling filters, elutriators); (B) Industrial water and waste treatment, e.g., coal washer refuse, steel mill flue dust washer thickeners, metal plating wastes, refinery wastes, pulp and paper mill effluents, water from many chemical processes	Sludge dewatering (sand drying beds, centrifuges, vacuum filters); Thickening and dewatering of wastewater solids; Coal preparation plants (fine coal slurry dewatering, refuse sludge dewatering) (vacuum filter or centrifuges)
600	Coagulant and/or coagulant aid in water clarification (0.5–5.0 ppm, and/or 0.5–2.0 ppm); In-line clarification of water (1–5 ppm); Clarification of surface water, process water (paper mill white water, etc.), cooling water (hot rolling mill cooling water in steel mill etc.)	(A) Municipal sewage clarification; (B) Industrial wastewater clarification, e.g., refinery and steel mill wastewaters	
(special) 600	Clarification of industrial and municipal water: coagulant (0.5–10 ppm), coagulant aid in conjunction with lime, alum, etc. (0.2–5 ppm)		
633			In pulp and papermaking: paper machine retention of fines fiber, filler materials (45–270 g/ton of fiber), save-all flocculation—solids recovery, cleaner white water for mill reuse (0.5–4.0 mg based on total influent), liquor clarification (white liquors: 5–10 mg, green liquors: 1–5 mg, calcium hypochlorite bleach liquor: 5–15 mg)

Organic Flocculant Trade Name	Public and Industrial Water Treatment	Application	
		Municipal (A) and Industrial (B) Wastewaters	Sludge Treatment and Industrial Suspensions
NALCO 634			In the manufacture of paper and paperboard (promoting the effectiveness of standard gum additives)
635			Pulp and papermaking: paper machine retention of fines, pigments and filler material (45–225 g/ton of dry fiber), save-all flocculation (0.4–2.0 mg, based on total influent), liquor clarification (white liquors: 5–10 mg, green liquors: 1–5 mg, calcium hypochlorite bleach liquors: 1–5 mg),
636			Papermaking: paper machine retention of fines and filler material ($\sim$22–225 g/ton), save-all recovery (0.2–2.0 mg, based on total effluent)
7717		(B) Breaking of oil-in-water emulsions	
7723		(B) Breaking and separation of oil-in-water emulsions	
NALCOLYTE 110	Coagulant aid in clarification of public water supplies and flocculant in softening of water (0.5–5.0 ppm); Clarification of high turbidity water (alone as flocculant); Clarification of highly colored low turbidity water (together with alum)	(A) Flocculant in clarification of municipal wastewaters; (B) Clarification (flocculant) of blast furnace gas wash water, white water systems and industrial wastewaters	Slurry flocculant, e.g., ore slurries, chemical slurries (sedimentation, 22–1,400 g per ton of dry solids)
603		(A) Clarification of municipal sewage wastewater in primary and secondary settling (1–10 ppm, may be used in conjunction with Nalcolyte 675, dosage 1–5 ppm)	Sewage sludge filtration (filter aid) (2–20 kg/ton of dry solids)
607	Coagulant (1.0–5.0 ppm) or coagulant aid in clarification of nonpotable water (up-flow sludge blanket and sludge recirculation type clarifiers); Precipitation softening processes		

Organic Flocculant Trade Name	Public and Industrial Water Treatment	Application	
		Municipal (A) and Industrial (B) Wastewaters	Sludge Treatment and Industrial Suspensions
NALCOLYTE 610		(A) Primary settlers (0.1–3.0 ppm)	Sludge dewatering (vacuum filters, solid bowl centrifuges, sludge drying beds, dosage in low solids sludges 1.2–3.2 kg/ton of dry solids, in high solids sludges 0.22–1.3 kg/ton of dry solids); sludge thickeners and flotation concentrates (225–900 g/ton of dry solids)
670	Industrial (nonpotable) supply clarification (0.01–1.0 ppm)	(B) Wastewater clarification, e.g., blast furnace and oxygen furnace flue dust washer (thickening and dewatering)	Thickening and dewatering of mineral and chemical process slurries: uranium milling (acid leach—countercurrent decantation, alkaline leach—ball mill thickeners), copper concentrates and tailings, potash tailings (4–450 g/ton of dry solids); Thickening and dewatering of sludges from blast furnace gas washes; Clarification of K_2CO_3 process solutions
671	Coagulant aid for improving clarification of turbid waters (0.1–1.0 ppm); Filter aid (with alum) in in-line clarification of turbid waters (0.005–0.05 ppm); Precipitation softening coagulant aid (0.1–1.0 ppm)		
672	Industrial (nonpotable) supply clarification (0.01–1.0 ppm)	(B) Waste wash clarification—blast furnace and oxygen furnace flue dust thickening and dewatering, rolling mill scale pit slurry clarification	Separation of chemical process slurries (settling, filtration systems, 4–450 g/ton of dry solids), conditioning of sewage sludges
673		(B) Industrial wastewater and thin slurries clarification (0.01–1.0 ppm), clarification of recycle wash water (coal preparation)	Separation of chemical process slurries (4–450 g/ton of dry solids); settling, thickening and dewatering (by filtration or centrifuging) of coal fines and refuse solids; sludge dewatering from KALDO-furnace process
675		Municipal and industrial wastewater, thin slurry clarification (0.01–	Industrial and municipal sludges dewatering, separation of chemical process slur-

Organic Flocculant Trade Name	Public and Industrial Water Treatment	---------- Application ---------- Municipal (A) and Industrial (B) Wastewaters	Sludge Treatment and Industrial Suspensions
NALCOLYTE 675		1.0 ppm)	ries (settling and filtration systems, e.g., phosphoric acid clarification, 4–450 g/ton of dry solids); taconite tailings clarification
8101	Clarification and sedimentation of flocs in nonpotable water treatment		
D 1782	Treatment of industrial water	Treatment of industrial wastes	Flocculation of coal and other mines slurries
PEI 1090		Flocculation of algae	
POLYFLOK (generally)	Coagulation, sedimentation and filtration aids in the treatment of raw water (0.05–1.0 ppm); Filter aids (0.01–0.1 ppm) (horizontal or upward flow clarification tanks)	Clarification of filter backwash water (1,000–5,000 mg of a powder Polyflok grade); (A) Sewage treatment (primary sedimentation aid, 2–10 ppm of powder Polyflok grade; sedimentation aid for humus sludge 2–10 ppm); (B) Clarification of coal washery effluents	Certain grades as filter aids (drum or disc filters, press plate filters) or organic conditioners of sludges or slurries (e.g., sewage, coal slurry, paper, metallic hydroxides, etc.); Dewatering (sludge conditioning aids) of raw or digested sewage sludges on drying beds (raw sludges 1,500 mg, digested sludges ~3,000 mg of powder grade Polyflok on dry solids), vacuum filter (raw sludge 1,500 mg, digested sludge 3,000 mg powder grade Polyflok on dry solids), filter press (1,500 mg raw sludges, 3,000 mg digested sludge, of a powder grade Polyflok, based on dry solids), solid-bowl centrifuges (1,500–4,000 mg of a powder Polyflok grade, based on dry solids); Sewage sludge consolidation prior to the disposal (500–3,000 mg of powder Polyflok grade, based on dry solids); For improving the settling rate of suspended solids, settling lagoons of a sand and gravel quarry, conventional thickeners such as are used in mining and similar industries (0.01–0.05 kg/ton of dry solids); Sedimentation aid (thickening) for slurry, e.g., in the coal mining industry, and

Organic Flocculant Trade Name	Public and Industrial Water Treatment	Application — Municipal (A) and Industrial (B) Wastewaters	Sludge Treatment and Industrial Suspensions
POLYFLOK			tailings from froth flotation plants (0.01–0.1 kg/ton of dry solids); Filtration aid of frothed concentrates and thickener underflow prior to vacuum filtration (0.01–0.1 kg/ton of dry solids); Flotation in air-flotation units (e.g., save-alls for fiber recovery in paper industry, thickening of surplus activated sewage sludge, separation of fats in food industry, etc.; Sludge conditioners prior to filter pressing (0.1–0.2 kg/ton of dry solids); Centrifugation/dewatering, e.g., in tailings and fine coal, sewage sludge, water softening sludges (0.2–0.5 kg/ton of dry solids); Sludge consolidation from the sedimentation stage in a water treatment process (1,000–5,000 mg of powder Polyflok, based on dry solids); Dewatering of hydroxide sludges from a water treatment process (drying beds: 1,000–5,000 mg, vacuum filters: 1,000–5,000 mg, filter press: 2,000–5,000 mg of a powder Polyflock grade, based on dry solids)
POLYFLOK 4D, 8H, 100X			Use in mineral dressing processes, in coal industry, thickening of tailings; As filter aid (drum or disc filters, press plate filters)
37CP		(A) Sewage treatment, solids removal in primary sedimentation tanks (5–10 ppm), to improve the separation in either settlement or flotation units of the solids in secondary treatment processes (4.5–6.8 g/ton of dry solids)	Conditioning of raw and digested sludge prior to dewatering by mechanical means (11–18 kg/ton of dry solids)

Organic Flocculant Trade Name	Public and Industrial Water Treatment	Municipal (A) and Industrial (B) Wastewaters	Sludge Treatment and Industrial Suspensions
		--------- Application ---------	
POLYFLOK 8039 PR			Dewatering in paper industry
PX			Improving the settlement rate of tailings (~90 g/ton of suspended solids)
POLYHALL M 60			Improvement of digested sludge filtration (0.5–2.0% of dry solids)
402			Flocculation of diluted suspensions (0.025–5.0 ppm), thickening of suspensions (2.2–226 g/ton of dry solids), improvement of suspension filterability (4.5–226 g/ton of dry solids)—especially effective on acid leached uranium slurries, acid leached copper ores, base metal concentrates, copper hydroxide precipitates, etc.
M-295			Clarification of dilute slurries (0.025–2.5 ppm), thickening (2.2–226 g/ton of dry solids), filter aid (4.5–226 g/ton of dry solids), industrial suspensions (especially effective on coal flotation residue, gold tailings, base metal tailings and concentrates, certain iron ore slimes, magnesium hydroxide slurries, etc.)
M-630		Clarifying aid used with Polyhall M-295	Filter aid for domestic sewage digested sludge (0.5–2.0% based on dry solids)
POLYMER X-150	See Polyox MN		Flocculation and sludge dewatering in centrifuges; Coal slurries and potash slime by centrifuging (5 g/m^3) Cooling system defouling agent (3 ppm)
225, 226, 233, 243, 253	See Calgon Coagulant Aid		
X 1633		(B) Clarification of wastewater from brass production	
POLYMIN HS, KM, P, SW			Dewatering in paper industry

--Application--

Organic Flocculant Trade Name	Public and Industrial Water Treatment	Municipal (A) and Industrial (B) Wastewaters	Sludge Treatment and Industrial Suspensions
POLYOX			Clarification of coal slurries, coal washes, selective flocculant in Mn ore processing
MN Coagulant	Flocculation of river water turbidity ($<$1 ppm)	Flocculation and separation of mineral suspension (sedimentation, centrifuging, $<$5 g/m^3)	Flocculation and dewatering of sludge in centrifuges; Coal slurries and potash slime (5 g/m^3); Flocculation and centrifuging in systems involving clays, coal tailings, dust, silica, metals, potash and other minerals
POLYTERIC AS 3			Clarification of wastewaters from Cu mining
POWDAFLOK			Clarification of coal, gravel washing and mineral ore separation plant effluents
PRAESTOL (generally)	Treatment of surface water (in clarifiers) for potable and nonpotable use; Decarbonization of water in clarifiers or settling tanks; Sand filter backwash treatment; Treatment of water for operational purposes	(A) Clarification of water from digestion towers, improved sedimentation in mechanical pretreatment; (B) Stone coal industry: flocculation of suspensions in coal washes; Stone and clay industry: flocculation and sedimentation of sand and gravel washes; Paper industry: purification of papermaking machine effluents (sedimentation and flotation), paper mill effluents; Metallurgy: flocculation of top gas and converter gas washes, treatment of cooling waters from rolling mills, flocculation of hydroxide flocs in neutralizing water from galvanizing plants and pickling plants; Chemical industry: purification of wastewaters with oil content (with inorganic salts), purification of recirculation acid liquids (pH 1,	Various mechanical sludge dewatering—vacuum filters, centrifuges; Hydroxide sludge from water treatment, clay sludges (solid-bowl screw centrifuge); Inorganic mineral suspensions—vacuum filters, drum and disc centrifuges (anionic polymers); Predominantly organic sludges—vacuum filters, solid-bowl, screw centrifuges, screen strip filter presses (cationic polymers); Dewatering of sludges from mechanical and biological wastewater treatment plants—solid-bowl screw centrifuge, screen strip filter presses; Municipal wastewaters: sludge concentration (in biological treatment), raw sludge dewatering (vacuum rotary filters, vacuum belt filters, screen belt filter presses, solid-bowl screw centrifuges, drying beds); Stone coal industry: dewatering of raw sludges and flotation concentrates

Organic Flocculant Trade Name	Public and Industrial Water Treatment	Application— Municipal (A) and Industrial (B) Wastewaters	Sludge Treatment and Industrial Suspensions
PRAESTOL		phosphoric acid, sulfuric acid, ilmenite leach)	(vacuum filters), mechanical dewatering of concentrated flotation wastes (solid-bowl screw centrifuges); Ore processing industry: thickening of crushed ores for flotation; flocculation and sedimentation of flotation wastes, mechanical dewatering of flotation concentrates (vacuum filters), of flotation wastes; thickening and dewatering of concentrates and wastes from wet and magnetic treatment; filtration and clarification of spent liquor from chemical processing of ores; Potash industry: sedimentation of suspension in solvent liquor; purification of lye after decarbonization; dewatering of effluents after purification and decarbonization (solid-bowl screw centrifuge); Stone and mineral industry: dewatering of sludges from treatment and thickening of raw material (solid-bowl screw centrifuge, filters); treatment of kaolin suspension, reclamation of kaolin before dewatering in filter presses; Paper industry: retention aid, increased dewatering at screen stage, mechanical dewatering of sludge from effluent; Metallurgy: dewatering of sludge from effluent (vacuum filters, solid-bowl centrifuges, screen belt filter presses); Chemical industry: dewatering of sludges from production processes (acid leach, vacuum filters); clarification—washing and filtration of suspensions in pigment and dyestuff production; dewatering of sludge from mechanical and biological purification of wastewaters (screen belt filter presses, solid-bowl screw centrifuges)

Organic Flocculant Trade Name	Application		
	Public and Industrial Water Treatment	Municipal (A) and Industrial (B) Wastewaters	Sludge Treatment and Industrial Suspensions
PRAESTOL 114K	Improved flocculation (0.04–1.0 g per cubic meter)		
114K, 184K			Dewatering of fine suspensions (filtration-filter presses); Retention aid in paper production (1 kg per ton of solids)
222K			Retention aid in paper production
444K, 500K		(B) Papermaking machine effluents (0.4–1.5 g/m^3)	Dewatering of sludge from municipal sewage treatment plants (solid-bowl turbine centrifuge); retention aid in paper production
2450	Flocculation of coagulate turbidity particles (0.3–3.0 g/m^3)		Dewatering of industrial suspensions (concentrates in ore processing, steel and iron industry, stone industry)—vacuum filters, solid-bowl centrifuges (10–30 g per ton of dry solids)
2700, 2750	Sedimentation of turbidity particles (0.2–2.0 g/m^3)	Sedimentation of suspended matter (2–8 g/m^3)	Dewatering of sludge and/or crushed ores (solid-bowl screw centrifuges)—150–300 g/ton of dry solids), and/or decantation centrifuges
2800	Coagulation aid (0.04–1.0 g/m^3)	Flocculation of suspensions (2–8 g/m^3)	Dewatering of sludges in solid-bowl screw centrifuges; pressure filtration of hydrophylic kaolin and bitumen suspensions; Flocculation of effluent suspensions from iron and steel processing; stone and mineral industry; Paper production: retention aid, and/or dewatering on paper machine screen
2820, 2830	Coagulant aid (0.1–1.0 g/m^3)	Flocculation of suspensions (2–6 g/m^3); Wastewater from paper machine (0.4–1.5 g/m^3) (Praestol 2820)	Improving of suspension sedimentation in acid medium) dewatering of suspensions and sludges (solid-bowl screw centrifuges, 150–400 g/ton of dry solids)
2850		Improving of suspension sedimentation in water (1.0–4.0 g/m^3)	Dewatering of flotation waste suspensions by filtration (filter presses); Flocculation of effluent suspension in iron and steel processing; stone and mineral industry;

Organic Flocculant Trade Name	Public and Industrial Water Treatment	Municipal (A) and Industrial (B) Wastewaters	Sludge Treatment and Industrial Suspensions
		------Application------	
PRAESTOL 2850			Potash production: clarification of potash lye (0.1–0.4 g/m^3), centrifuging of suspension K_2SO_4 (1.5 g/m^3)
2900		Improving of suspension sedimentation in neutral medium (1.0–4.0 g/m^3); Wastewaters from paper machine (0.4–1.5 g/m^3)	Dewatering of fine suspensions in solid-bowl screw centrifuge (e.g., sludges from paper machine wastewaters, 20–100 g/m^3); Flocculation of effluent suspension in iron and steel processing; stone and mineral industry
2935		Paper machine wastewaters (0.4–1.5 g/m^3)	Dewatering of industrial slurries and sludges from wastewater purification (e.g., paper wastewaters–solid-bowl screw centrifuge, 20–100 g/m^3), from KALDO-furnace process
3000			Paper industry: retention aid and/or dewatering on paper machine screen; Dewatering of sludges in solid-bowl screw centrifuge
FZN			Clarification of sugar juice
PRIMAFLOC (generally)	Prime coagulant–clarification of highly and moderately turbid waters; Clarification of slightly turbid waters as prime coagulant in conjunction with a suitable clay (bentonite); Coagulant aid with inorganic coagulant	(A)(B) Treatment of both industrial and domestic wastewaters	Liquid-solids separation as part of a manufacturing process
C-3	As a prime coagulant (turbid waters, 1–10 ppm) or as coagulant aid for clarification of water (without or in combination with clay–slightly turbid water)	Certain waste treatment applications	Conditioning of industrial sludges and wastes streams prior to mechanical dewatering (vacuum filtration, flotation or centrifugation); Conditioning of fermentation broths
C-5	Prime coagulant in raw water clarification in conjunction with clay (3–15 ppm);	Certain waste treatment applications	

Organic Flocculant Trade Name	Application		
	Public and Industrial Water Treatment	Municipal (A) and Industrial (B) Wastewaters	Sludge Treatment and Industrial Suspensions
PRIMAFLOC C-5	Coagulant aid after alum-inorganic coagulant with the addition of clay following the polymer (0.3–2.0 ppm)		
C-6	As prime coagulant in clarifying moderate or high turbidity waters (1–10 ppm), in low turbidity water the addition of clay (bentonitic); as a coagulant aid with an inorganic coagulant (0.1–1.0 ppm)	Certain waste treatment applications	Conditioning of domestic and industrial waste streams prior to mechanical dewatering (vacuum filtration, flotation, centrifugation, 0.1–1.0% of dry solids)
C-7	Prime coagulant in flocculation of solid particles in industrial raw water with moderate and high turbidity; coagulant aid for slight turbidity water (suspension below 50 ppm), used together with clay, and/or inorganic coagulants (1–10 ppm)	(A) Clarification of raw sewage together with Primafloc A-10 (typical dosage: C-7, 0.1–0.5 ppm; A-10, 1–5 ppm) (B) Waste treatment application in industry); Flocculation of wastewater with resins emulsion content	Dewatering of sludges and slurries (rotary vacuum filters—0.1–1% of dry solids, solid-bowl centrifuges, sludge drying beds); Thickening (concentration) of waste activated sludge using air flotation (dosage 500–4,500 g/ton of solids); Industrial solid-liquid separations (in ceramic, photographic, mining, paper, cement, coal, fermentation and fiber industries)
A-10		(B) Purification of wastewaters with lime content from the production of sole leather	Improving of suspensions separation-filtration, centrifuging, flotation or sedimentation (certain organic and inorganic suspensions in industrial processes, industrial waste streams)
XH-10			Conditioning suspended particles for mechanical separation techniques—filtration, centrifugation, flotation
PROSEDIM			Dewatering of sludge by filtration
PURIFLOC (generally)		Clarification of raw sewage	Conditioning of raw sewage sludge for vacuum filtration; Thickening of suspensions by sedimentation, flotation, centrifuging; Dewatering of sludges—vacuum filters, centrifuges, drying sand beds

Organic Flocculant Trade Name	Public and Industrial Water Treatment	Application — Municipal (A) and Industrial (B) Wastewaters	Sludge Treatment and Industrial Suspensions
PURIFLOC (generally)			Municipal sewage treatment works: thickening and dewatering of sludges—gravity flotation, centrifuging, vacuum filters, sludge drying beds, elutriation
(anionic)	Paper industry: clarification of water for kraft pulp mill ($\sim$0.5 ppm), for paper mill—fine paper ($\sim$0.6 ppm), paper mill (as aid to the alum)	Paper industry: clarification of wastewater from kraft pulp mill ($\sim$1.5 ppm), paper mill fine paper ($\sim$1.5 ppm), paper mill—box board ($\sim$0.5 ppm, in air flotation save-all); Metal finishing operations—automotive: clarification of electroplating wastewater from automotive parts plant (increase the oil removal, flotation unit); from automotive assembly plant (with $FeCl_3$), from electronic plant; Food industry: clarification of wastewater from meat packing plant (air flotation unit), from cereal mill (sedimentation unit), from vegetable oil processing plant (air flotation unit); Oil production and oil refining operations: clarifying the general refinery wastewater in sedimentation units; Tanning industry: clarification of wastewater in an air flotation unit or sedimentation basins, tannery wastewater with a small amount of chrome (primary sedimentation, reduction of suspended solids), treatment of waste from sheep and skin side leather processing (primary sedimentation units); Fiber glass: wastewater from the manufacture of glass fiber (pri-	Metallurgy: thickening of sludge from wastewater purification plants (2 ppm); Automobile industry: dewatering of suspensions; Vacuum filters; Dewatering of sludge from fly ash water treatment (2 ppm); Glass fiber production; Dewatering of sludge from the treatment of wastewater from synthetic fiber production (Zn content)

Organic Flocculant Trade Name	Application		
	Public and Industrial Water Treatment	Municipal (A) and Industrial (B) Wastewaters	Sludge Treatment and Industrial Suspensions
PURIFLOC (anionic)		mary clarifier); Textile industry: clarification of wastewater from production of synthetic fibers (sedimentation basin); Steel industry: clarification of steel mill-flue dust wastewater, of steel mill-rolling mill wastewater; Alumina industry: clarification of aluminum mill-reduction furnace wastewater, of aluminum mill-rolling mill wastewater	Paper industry: sludge conditioning (22–45 g/ton of dry solids); Brewery: thickening of waste activated sludge in an air flotation unit (4.5–5 kg per ton of dry solids), sludge dewatering from brewery, biological waste treatment plant, conditioning of sludge prior to sand bed dewatering; Cereal mill: conditioning of primary and waste activated sludge prior to air flotation thickening, and/or to vacuum dewatering; Milk industry: conditioning of sludge in activated sludge treatment plants; Textile industry: dewatering of digested sludges from biological treatment of textile wastes (on vacuum filters); Reduction of sludge volume by biological treatment of wastewater (8 ppm), dewatering of sludge (sand bed, 2.0 kg per ton); Dewatering of sludge from biological treatment of wastewater (vacuum filters, 11 kg/ton of dry solids)
(cationic)	Paper industry: water clarification for deinking mill (as aid in an upflow clarifier); Chemical industry: water clarification, as a filter aid in an upflow filter	Food industry: clarification of wastewater from meat packing plant, from milk processing plants (activated sludge); Brewery: clarification of wastewater in an activated sludge treatment plant; Cereal mill: clarification of wastewater (bulking sludge at the activated sludge treatment plant); Oil production and oil refining operations: clarification of wastewater in an air flotation unit and API separator; Glass: clarification of glass-polishing waste in wastewater lagoons; Ceramics: clarification of wastewater from ceramic tile production; Textile industry: clarification of wash water in a cotton pulping operation; Soap industry: clarification of the effluent from air flotation units	
(nonionic)	Paper industry: water clarification	Paper industry: wastewater clar-	Chemical industry: dewatering of various

Organic Flocculant Trade Name	Application		
	Public and Industrial Water Treatment	Municipal (A) and Industrial (B) Wastewaters	Sludge Treatment and Industrial Suspensions
PURIFLOC (nonionic)	for pulp mill (with alum, in upflow solids-contact clarifier); Aluminum industry: water clarification for aluminum processing and fabrication plant (as an aid to the inorganic flocculants); Oil refinery: water clarification for heat exchanger (scaling)—sedimentation basin	ification from deinking mill; Metal finishing operations—automotive: clarification of wastewater from automotive parts plant; Oil production and oil refining operations: wastewater clarification (clarifying an API separator effluent in gravity units), clarification of general refinery wastewater (with alum)	waste sludges in centrifuge units (2.7 kg/ton of dry solids)
(cationic + anionic)		Paper industry: clarification of wastewater from sulfite pulp mill ($\sim$5 ppm cationic + $\sim$2 ppm anionic); Phosphorus manufacture: clarification of wastewaters from phosphate processing (sedimentation basins)	Vegetable oil processing plant: dewatering of sludge from wastewater clarification (vacuum filter, 2 ppm + 20 ppm); Fiber glass: conditioning of sludge from primary and secondary clarifiers prior to dewatering on vacuum filters
A 21		(A) Flocculation of raw sewage; (B) Clarification of meat packing wastewater; Improving efficiency of trickling filter plants	Thickening of sludge from primary sedimentation (municipal sewage purification plants)
A 22		Clarification (and thickening) of flue dust wastewaters	Thickening of sludge from electroplating (hydroxide sludges)
C 31	Flocculation of industrial raw water, raw river water	Flocculant in systems: industrial and municipal wastewater; Clarification of oily wastewater, paper mill wastewater; Improving efficiency of activated sludge plants	Thickening of sludge—gravitation, centrifuging, flotation; Dewatering of municipal and industrial sludge—vacuum filters, centrifuges, sand beds; Activated sludge treatment; Flocculation of algae
601			Flocculation of bulked activated sludge
602			Flocculation of bulked activated sludge

Organic Flocculant Trade Name	Public and Industrial Water Treatment	Municipal (A) and Industrial (B) Wastewaters	Sludge Treatment and Industrial Suspensions
RETEN (generally)		(B) Industrial wastes and process streams: for example, organo-metallic waste stream ~10 ppm; Paper mill waste stream (in conjunction with alum); Metal plating waste stream (Ni)	Secondary treatment: sludge (raw and digested) conditioning for all types of dewatering equipment (vacuum filtration), dewatering partially digested sludge in the Roto-Plug, a Bird centrifuge; An aid to settling sewage solids in overloaded elutriation systems, in activated sludge processes; Sedimentation aid; An aid to pressure flotation; Industrial liquid-solids separations; Cement plant—dust reclamation circuit (flocculant for Portland cement plant flue dust); Flocculation of iron ore flotation tailings stream; Paper mills—retention aid; Concentrating numerous process streams (as coal fines, cement, uranium ores); Metallurgy: filter aid for material from leached solutions; Selective flocculation
205	Coagulant aid (0.15–0.25 ppm)	Waste streams from a variety of industrial operations (especially at pH 6–10); (B) Clarification of wastewater from nickel plating—conditioning of neutralization sludge (0.4–1.0 kg/ton of dry solids)	Dewatering of raw and digested sludges (vacuum filters ~900 g/ton of dry solids), treatment of sludge by elutriation (~450 g/ton of dry solids), dewatering of partially digested sludge (Roto-Plug centrifuge), treatment of activated sludge by flotation; Retention aid in papermaking (Reten 205M, 205 MH ~45–225 g/ton)
210			Dewatering of sludge on vacuum filters
SEDIPUR (generally)	Clarification of raw river water (in conjunction with inorganic coagulants); Filter aid (<1 g/m³)	Industrial wastewaters: from brown coal processing, rock oil refinery, Pb-Zn ores processing, Fe ores, limestone, porphyrin,	Dewatering of thickened flotation wastes (centrifuges), dewatering of hydroxide sludges (filtration, centrifuging—2–10 g per cubic meter of sludge);

Organic Flocculant Trade Name	Public and Industrial Water Treatment	Application Municipal (A) and Industrial (B) Wastewaters	Sludge Treatment and Industrial Suspensions
SEDIPUR (generally)	Decarbonization of water by lime (<1 g/m^3);	from porcelain production, ceramic industry, from gas works, metallurgical plants, galvanization plants, pickling plants, tanning industry, stone coal—treatment of recirculation water; sugar production—clarification of wash water	Thickening of suspensions by sedimentation (2–10 g/m^3 of sludge); dewatering of settled sludge (filtration, centrifuges); thickening and dewatering in coal industry (filtration, centrifuges); Coal: flocculation of flotation tailings, filtration aid; Ore industry: Pb-Zn ores—thickening of concentrates, flocculation of flotation tailings, improvement of filtration; U ores: treatment—sedimentation of extraction residue; Cu ores: sedimentation of residue, thickening of concentrates; rare metals: sedimentation of residue, thickening of concentrates (cyanide leaching); Bi gaining—sedimentation of Bi chloride and its filtration; Ni and Co gaining—flocculation of hydroxide precipitates; kaolin—flocculation of kaolin suspension; salt gaining—clarification of NaCl solution; sugar production—clarification of sugar juice
TF-fest			Dewatering of sedimentation sludge from rolling mills, pickling plants, galvanization plants (vacuum filters, pressure filters, centrifuges—5–15 g/m^3 of sludge); sedimentation separation of insoluble residue after diluted H$_2$SO$_4$ leaching of U ore ($\sim$50 g/ton of ore), thickening by sedimentation of precipitated hydroxides of Co and/or Ni by their gaining from ores ($\sim$20 g/l), thickening (1–5 g per cubic meter) or improving of filtration characteristics (2–10 g/m^3) by neutralization of resulting waste suspensions by Zn ore processing; Pressure filtration of hydrophilic kaolin and bitumen coal suspensions

Organic Flocculant Trade Name	Application		
	Public and Industrial Water Treatment	Municipal (A) and Industrial (B) Wastewaters	Sludge Treatment and Industrial Suspensions
SEDIPUR T1-fest		(B) Clarification of wastewater by flocculation and sedimentation—rolling mills (0.1–2.0 g/m^3), neutralized waters from pickling plants (0.1–5.0 g/m^3), galvanization plants (0.1–2.0 g/m^3); flocculation of wastewater from refineries (oil content at least 100 mg) and treated with ferrous sulfate and lime (0.5–2.0 g/m^3)	Separation of concentrated suspensions by filtration or centrifuging (5–20 g/m^3); clarification of sugar juice (0.1–5.0 g/m^3); Conditioning of sludge from municipal sewage plants (with cationic polymers)
TF 2-fest		(B) Flocculation of wastewaters from refineries (oil content at least 100 mg) treated with ferrous sulfate and lime (0.5–2.0 g/m^3)	Clarification of sugar juice (0.1–0.5 g/m^3)
TF 2	In-line clarification of surface waters (0.1–1.0 g/m^3)		
TF 5-fest		(B) Flocculation of wastewaters from refineries (oil content at least 100 mg) treated with ferrous sulfate and lime (0.5–2.0 g/m^3); flocculation (sedimentation) of neutralized wastewater (suspension of metal hydrolysis) ($\sim$5 g/m^3)	Thickening of inorganic sludge (5–20 g/m^3)
TF 5		(B) Flocculation of wastewaters from tanneries with Cr-salt content (together with FeSO$_4$); Clarification of wastewater from sugar beet washing (sedimentation $\sim$1 g/m^3)	Pressure filtration of hydrophilic kaolin and bitumen coal suspensions; Separation of kaolin from its aqueous suspension by sedimentation ($\sim$0.5 g per cubic meter); clarification of brine (NaCl) (0.1–1.0 g/m^3); Dewatering of sludge from KALDO-furnace process; Clarification of suspended salts solutions (0.1–1.0 g/m^3)
KA		(A) Clarification of municipal sewage by flocculation and sedimentation (0.3–15.0 g/m^3 of turbid	Thickening and dewatering of organic sludge from municipal sewage plants—filtration of raw and digested sludge

Organic Flocculant Trade Name	Public and Industrial Water Treatment	---Application--- Municipal (A) and Industrial (B) Wastewaters	Sludge Treatment and Industrial Suspensions
SEDIPUR KA		matter), by clarification with activated sludge (sedimentation); (B) Clarification of wastewater from sand washing, waters with aluminum silicate content, waters from dust separator from convertor	(vacuum filters: 300–1,000 g/m^3), digested sludge (sludge beds, centrifuges: 300–1,000 g/m^3), sedimentation of activated sludge (0.3–15.0 g/m^3); Flocculation of inorganic suspensions; Clarification of mine waters; Clarification of flotation tailings
KA + T1, and/or TF 5		(A) Primary sedimentation of inorganic and organic suspension mixture in municipal wastewaters (dosage of KA ~5–20 g/m^3)	Thickening and dewatering of inorganic and organic sludge mixture by filtration (ratio of T 1, and/or T 5 to KA of 1:10 to 1:1); Clarification of flotation tailings
SEDOSAN		(B) Clarification of wastewaters from papermaking (in conjunction with Al^{+++} salts)	
SEPARAN (generally)	Clarification of water for industrial and potable use (coagulant aid, NP-10, AP-30) Primary flocculants in the treatment of raw river water where the solids level is high (NP-10, AP-30: 0.01–2.0 ppm)	(B) Industrial waste treatment, e.g.,: flue dust recovery, electroplating wastes, oil well injection water—secondary oil recovery (as secondary coagulant when used with lime and ferric sulfate, reduction in effluent turbidity—dosage ~0.15 ppm)	Mining and ore processing; Other applications in the production of ammonia, barium, brines (e.g., clarification of saturated KCl-NaCl brines), clay, drugs, fine chemicals, inorganic salts, lithium, soda ash, pigment
C-120		Clarification of wastewaters	Thickening and dewatering of sludge
AP-30	Flocculation of river water with 1–15% suspension content (0.01–2.0 ppm)	(A) Flocculation of municipal wastewaters with Al^{+++} salts	Selective flocculation in ore industry, sugar juice clarification; filtration of sewage sludge; Coal washeries and settling and filtration of coal fines and slimes; Magnesia production: treatment of thickener feed streams (~9 g/ton of solids)
AP-273			Sugar juice clarification
NP-10	Flocculation of river water with 1–15% suspension content (0.01–2.0 ppm);	(A) Flocculation of municipal wastewaters with Al^{+++} salts; (B) Clarification of industrial waste-	Flocculation of digested elutriated sewage sludge prior to dewatering on vacuum filters (~230–450 g/ton of solids);

Organic Flocculant Trade Name	Public and Industrial Water Treatment	Application Municipal (A) and Industrial (B) Wastewaters	Sludge Treatment and Industrial Suspensions
SEPARAN NP-10	Hot and cold lime softening	water (0.01–5.0 ppm); Clarification of petroleum refinery oily wastes in flotation units in conjunction with alum (0.5–2.0 ppm); Clarification of industrial wastewater by sedimentation (0.01–5.0 ppm); Removal of suspension of recirculation water by deep drilling (sedimentation 3–15 ppm); Flocculation of paper mill wastewater (white liquor, green liquor–~0.25–5.0 ppm); Clarification of combined deinking and white water effluents wastes (~0.5–2.0 ppm), flocculation of wastewater from viscose production; Drilling water clarification (~3–15 ppm)	Flocculation of mine underground water; Mine and ore processing: Flocculation and settling of slime in concentration of low-grade iron ores (4–12 g/ton of ore); Cyanide leaching of gold ores (thickening of the ground ore pulp, liquids-solids separation by filtration or countercurrent decantation, 45 g/ton of ore); Uranium (thickening preleach ore slurries 2–9 g/ton; carbonate leach filtration 45–180 g/ton; acid leach filtration ~22–90 g/ton; countercurrent decantation thickeners, 9–45 g/ton); Zinc calcines (acid-leaching of flotation concentrate calcines in a countercurrent decantation operation 4–45 g/ton; slime removal in electrolytic purification steps); Copper oxide ores (clarification steps following the acid leach and the precipitation stages, 9–45 g/ton of solids); Thickening flotation Cu concentrates (2–18 g/ton); In electrolytic refining of copper and zinc; Tailings disposal (e.g., copper plant); Preliminary thickening in potash ore flotation circuit (9–22 g/ton of ore); Pressure filtration of kaolin, dewatering of radioactive sludge $Fe(OH)_3$ by decantation centrifuge Process industries: Alum: settling of slime in several aluminum sulfate production plants; Borax: settling of clay impurities from hot borax streams (~32 g/ton of solids); Cement: thickening in wet process cement manufacture (~10–30 g/ton of solids);

Organic Flocculant Trade Name	Application		
	Public and Industrial Water Treatment	Municipal (A) and Industrial (B) Wastewaters	Sludge Treatment and Industrial Suspensions
SEPARAN NP-10			Lime sulfur: filtration in the production of lime sulfur for agriculture purpose ($\sim$22–45 g/ton of solids); Phosphoric acids: settling of solids in green acid ($\sim$10 g/ton of gypsum); Inorganic fiber slurries: asbestos-cement industry; Pulp and paper industry: filler retention, flotation type save-all
MGL			Selective flocculation in ore industry, thickening of clayey and siliceous suspensions in hydrocyclone
2610			Pressure filtration of kaolin suspension
SUPERFLOC (generally)	Municipal and industrial water treatment: clarification of potable and industrial raw water (alone or in conjunction with inorganics); Potable water grade in the treatment of water for public supply, as clarification agents for water used to wash fruits and vegetables ($<$10 ppm), in the clarification of beet sugar and cane sugar juices ($<$5 ppm by weight of juice); Flocculants (0.2–2.0 mg), flocculating aids (0.05–1.0 mg), filtration aids (0.01–0.1 mg)	Municipal waste treatment: raw sewage settling, clarification of primary and secondary effluents, digested supernatant clarification, polishing process in tertiary treatment, filtration procedures, oily-waste clarification; Industrial waste treatment: clarification of wastewater from—metal finishing, pulps and paper manufacture, brewing, leather tanning soap and detergent manufacture, glass and ceramic production, synthetic rubber production, meat packing, oil refining, food processing, textile processing, chemical processing, aluminum-steel-iron processing, sugar refining, paint manufacture; Steelworks (waste pickle liquors), gas-washing operation; Removal of phosphates from sewage effluent (with alum or ferric salts) (anionic types)	Municipal waste treatment: sludge thickening by gravity, centrifugation, and air flotation; Sludge dewatering on drying beds, vacuum filters, centrifuges, filter presses; Sludge elutriation; Alum sludge dewatering; Control of sludge bulking; Industrial waste treating: dewatering sludges from industries; Sludge conditioning agents (10–200 mg); Conditioning of dewatered sludge prior to incineration

Organic Flocculant Trade Name	----------Application----------		
	Public and Industrial Water Treatment	Municipal (A) and Industrial (B) Wastewaters	Sludge Treatment and Industrial Suspensions
SUPERFLOC C 100 C 110	PWG—in the treatment of water for public supply	(A) Primary settling or clarification of secondary effluents (0.1–0.5 ppm)	Conditioning of industrial or municipal sludges prior to vacuum filtration or centrifugation (0.2–2.5 mg/ton of dry solids); dewatering of sewage sludges on drying beds (C 110); concentration of activated sludge by air flotation or gravity thickening (1–10 ppm)
N 100 N 100S	Flocculants (0.2–2.0 ppm), flocculant aids (0.05–1.0 ppm), filtration aids (0.01–0.1 ppm)		Sludge conditioning agents (10–100 ppm)
A 100, 110, 115, 125, 130, 137, 150	Flocculants (0.2–2.0 ppm), flocculant aids (0.05–1.0 ppm), filtration aids (0.01–0.1 ppm)	(A) Clarification of raw sewage (type A 150, 0.5–1.0 ppm); (B) Effluent from a food processing plant (with primary coagulant, type A 110)	Improvement of filtration characteristic of sludge from municipal waterworks (type A 100, dosage 10–100 ppm); Sugar clarification (types A 115, A 125, A 130, A 137, A 150, dosage 1–5 mg on the juice)
500 ser.			Dewatering sludges from municipal and industrial settling operations prior to filtration or centrifugation
521 (573, 577, 581)	Coagulant (0.2–1.0 and/or up to 10.0 ppm), and/or coagulant aid in water clarification (0.1–5.0 ppm); filtration aid (0.002–0.01 ppm); Clarification of industrial raw water supplies	(A) Clarification of secondary effluents (1–5 ppm), digester supernatant; (B) Demulsification of dispersed and emulsified oils and greases (2–10 ppm); Clarification of latex wastewaters (50–100 ppm)	For conditioning of wastewater sludges prior to vacuum filtration or centrifugation (4.5–45.3 and/or 5–50 kg/ton of dry solids); concentration of activated sludge by air flotation or gravity thickening (~10–40 ppm)
800 ser.	Industrial grades flocculant (0.2–2.0 ppm), flocculant aids (0.05–1.0 ppm), filtration aids in clarification of various industrial raw waters (0.01–0.1 ppm)	Flocculant, flocculant aid in clarification of various wastewaters, in-plant process water, inorganic wastes (particularly metal oxides)	Sludge conditioning agents (10–100 ppm); dewatering of sludge from municipal sewage plants (together with cationic polymers)
900 ser.	Industrial grades flocculant (0.2–1.0 ppm), flocculant aids (0.05–1.0 ppm), filtration aids (0.002–0.01 ppm); clarification of groundwater (alone or in conjunction with inorganic coagulants)	Waste treatment	Sludge conditioning agents (10–100 ppm); Industrial liquid/solid separation

Organic Flocculant Trade Name	Public and Industrial Water Treatment	-------- Application -------- Municipal (A) and Industrial (B) Wastewaters	Sludge Treatment and Industrial Suspensions
SURSOLAN P 5		Secondary treatment of wastewater in paper industry	Papermaking (retention of fibers and filler)
SWIFT'S POLYMER X 100, X 110, X 111, X 112	Dosage: Concentrated suspensions or sludges–0.01–1.0 lb/ton Low concentrated liquors–0.01–1.0 ppm by weight		Papermaking industry: retention aid by paper production; Acid leach in metallurgical industry; Organic sludges treatment
X 400, X 410, X 420	Dosage: Concentrated suspensions or sludges–0.01–1.0 lb/ton Low concentrated liquors–0.01–1.0 ppm by weight		Inorganic suspensions; Flocculation of mineral slurries, concentrates and waste rock; Special retention agent by titanium production
X 700	Flocculation of water for industrial purpose	(A) Flocculation of sewage; (B) Clarification of industrial wastewater	Separation in hydrometallurgy; Flocculation and dewatering of sludge
SYNTHOFLOC (generally)		(A) Flocculation of wastewater from municipal sewage plants; (B) Wastewaters from iron and steel works (blast furnace ash purification, granulation of slag); Wastewaters from rolling mills (sintering waters from rolling lines); Wastewaters from pickling plants (neutralization wastewaters, wastewaters from drawing plants, enamelling plants, galvanizing plants, etc.); Wastewaters from foundries (wastewaters from wet dust separation); Wastewaters from galvanization plants (neutralization wastewaters); Wastewaters from Al works (wastewaters from wet dust separation);	Flocculation of activated sludge; Dewatering of raw and digested sludge (centrifuges, filters, sludge drying beds); Dewatering of sludge from papermaking plants (centrifuges); Stone coal industry: sludge filtration, dewatering of sludge (filter presses, centrifuges); Potash production: clarification of circulation lye, dewatering of final product (centrifuges); Kaolin, baryte, magnesite processing, phosphate processing; Metal ores processing (flotation, magnetic wet separation, leach liquors, etc.); Hydrometallurgy: reclamation of Zn dust, clarification of zinc sulfate lye by the electrolysis of Zn, cementation of Cu (clarification of cementation lye, hydroxide precipitates treatment),

| Organic Flocculant Trade Name | ------Application------ | | |
	Public and Industrial Water Treatment	Municipal (A) and Industrial (B) Wastewaters	Sludge Treatment and Industrial Suspensions
SYNTHOFLOC		Wastewaters from nonferrous metals output (mine waters); Wastewaters from rock oil refineries (wastewaters with oil content, combination with $FeSO_4$ + CaO); Wastewaters from papermaking, paint, ceramic industry; Stone coal industry: clarification of wash waters, mine waters	electrolysis of alkali (wash water of raw salts); Chemical industry: clarification of spent solutions (Orr's white, TiO_2, phosphoric acid); Clarification of raw sugar juice
TOLFLOC 340 300	Flocculation of water for industrial purposes	(B) Flocculation of industrial wastewater, oily substances separation	Flocculation and dewatering of sludge
TR - Flocculants (generally)	Raw water treatment: as coagulation, sedimentation (0.05–1.0 ppm) and filtration aids (0.01–0.1 ppm); Flocculant of filter back wash water (1,000–5,000 ppm); Mining industry: water clarification	(A) Flocculation of sewage in primary sedimentation tank (2–10 ppm), in humus tank (2–10 ppm); (B) Iron and steel wastewater	Flocculation of raw or digested sewage sludges (~3,000 ppm); Dewatering of sewage sludges (vacuum filter: 1,500 ppm raw sludge, 3,000 ppm digested sludge; improved cake formation by filter press: 1,500 ppm raw sludge, 3,000 ppm digested sludge; solid-bowl centrifuges: 1,500–4,000 ppm; consolidation tank: 500–3,000 ppm); Sludge consolidation tank (sludge bleed) from water treatment (hydroxide sludge) (1,000–5,000 ppm, based on dry solids); dewatering of hydroxide sludges (drying bed 1,000–5,000 ppm, based on dry solids; vacuum filter 1,000–5,000 ppm, based on dry solids; filter press 2,000–5,000 ppm); Paper industry: retention aids; Mining industry: thickener slurry (coal) (0.01–0.1 kg/ton of dry solids); sedimentation of tailings from froth flotation plants; froth concentrates (vacuum filter 0.01–0.1 kg/ton of dry solids); filter press (0.1–0.2 kg/ton of dry solids);

Organic Flocculant Trade Name	Public and Industrial Water Treatment	Municipal (A) and Industrial (B) Wastewaters	Sludge Treatment and Industrial Suspensions
		--- Application ---	
TR - Flocculants			centrifuge (0.2–0.5 kg/ton of dry solids); increase of the rate of settlement of suspended solids in lagoons (0.01–0.05 kg per ton of dry solids); Red mud flocculation at alumina plants
25 AP			Improve the filtration characteristics of thickened inorganic sludges (especially in frothed coal filtration), especially rotary vacuum filtration units; Paper industry: save-all flotation (in the presence of alum, dosage 1–10 ppm)
90 AP	Coagulant aid in process water treatment (flocculation sand and clay suspensions	(B) Effluent treatment from paper industry (1–10 ppm)	Flocculation of shale tailings
91 AP, 93 AP	See 90 AP	(B) See 90 AP	See 90 AP Paper industry: dewatering of spent Al hydroxide sludge from process water treatment (drying beds, rotary vacuum filters, filter presses), save-all other type (1–10 ppm in the presence of alum)
95 AP	See 90 AP	(B) See 90 AP	Paper industry: save-all—other types (1–10 ppm in the presence of alum); Sugar industry: juice clarification (1–5 ppm)
11 APR			Retention applications in the paper industry at peroxide bleached pulps; retention of TiO_2
CLR			Dewatering aid for the paper industry (stable under chlorine conditions)
90 CP, 91 CP		(A) Sewage and other organic suspensions; (B) Effluent treatment from paper industry (1–10 ppm)	Flocculation of low pH systems sand and clay suspensions, other colloidal inorganic suspensions; Paper industry: save-all—other types (1–10 ppm)
11 CPR			Paper industry: retention aid, save-alls flotation (1–10 ppm)

Organic Flocculant Trade Name	Application		
	Public and Industrial Water Treatment	Municipal (A) and Industrial (B) Wastewaters	Sludge Treatment and Industrial Suspensions
TRAGAFLOK			Clarification of colliery and gravel washing effluents
UETIKON - Flocculants (generally)	By the treatment of water for potable and industrial use	(B) Clarification of wastewater from sand filter washing; Clarification of various industrial wastewaters, papermaking wastewaters (reclamation of paper fibers); Separation of oil emulsified in water	Improved sedimentation of suspension $(0.5–5 \text{ g/m}^3)$; Dewatering of sludge by filtration (50–500 g/m^3) and centrifugation (100–500 g/m^3)
VA-2			Clarification of sugar juice, clarification of recirculation water from sugar beet transport; Clarification of oxytetracycline solution, penicillin solution processing; Flocculation of wastes by apatite processing
WELGUM S	Clarification of back wash water (0.2–0.4 ppm); Flocculation aid in water treatment		Sludge treatment from water treatment plant (0.2–0.4 ppm)
WISPROFLOC (generally)	Treatment of water for potable use (sedimentation basins, clarifiers) (0.25–5.0 ppm); Treatment of boiler feeding water	Clarification of some wastewaters	
WT (Calgon Coagulant Aid) 2635	Primary coagulant, coagulant aid in conjunction with inorganic salts	(A)(B) Municipal and industrial wastewaters clarification	Separation of suspensions (sedimentation, centrifuging, flotation)
2640		(A)(B) Improved separation solids-liquid by clarification of municipal and industrial wastewaters	Conditioning of sludge from municipal and industrial wastewater clarification (filtration, sedimentation, flotation, centrifuging)
ZETAG 32		(B) Clarification of raw sewage	Thickening of raw sludge after clarification (mainly centrifuging)

Organic Flocculant Trade Name	Public and Industrial Water Treatment	Municipal (A) and Industrial (B) Wastewaters	Sludge Treatment and Industrial Suspensions
		----------Application----------	
ZETAG 51			Improvement of sludge characteristics and sludge thickening, preferentially digested sludges and sludges with lime content (flotation, vacuum filtration, coil filters, drying beds)
63			Dewatering of sludges (mainly secondary)— centrifugation, vacuum filtration, flotation, filter presses
92			Thickening and dewatering of sludges (centrifugation, flotation, coil filters, vacuum filtration); Sedimentation in primary and secondary stage of clarification
94			Thickening and dewatering of sludges (vacuum filtration, coil filters, flotation, drying beds)
ZT 650	Coagulant aid by clarification of river water with inorganic salts, by water softening	(B) Flocculation of industrial waste-waters	Thickening of suspensions from steel mills

Table 4

Manufacturers and Distributors
of Organic Flocculants

Number	Manufacturer and/or Distributor (D)	Product
1	Alginate Industries 22 Henrietta Street London W.C. 2, England	Welgum
2	Allyn Chemical Company 2224 Fairhill Road Cleveland, Ohio 44106 USA	Claron
2a	Allyn Chemical Co. Box 3040 Euclid, Ohio 44117 USA	Allstate
3	Allied Colloids Ltd. Cleckheaton Road, Low Moore Bradford, Yorkshire, England	Filtaflok Magnafloc Percol Polyflok Powdafloc Tragaflok Zetag
3a	Allied Colloids, Inc. 30 Church Street New York, N.Y. 10007 USA	Percol Vaptreat
3b	Allied Colloids, Inc. One Robinson Lane Ridgewood, New Jersey 07450 USA	Percol LT
4	Allstate Chemical Co. Box 3040 Euclid, Ohio 44117 USA	All-Flok-Ade Allstate
5	American Cyanamid Co. Water Treating Chemicals Dept. Berdan Ave. Wayne, New Jersey 07470 USA	Aerofloc Aerosol Cyanamid Magnifloc PAM Reagent X-309 Reagent 182 Reagent S-3000 Reagent S-3019 Reagent S-3171 Superfloc
5a	American Cyanamid Co. Industrial Chemicals and Plastics Division Wayne, New Jersey 07470 USA	Magnifloc Superfloc

Number	Manufacturer and/or Distributor (D)	Product
6	Borden Chemical 50 West Broad St. Columbus, Ohio 43215 USA	Agrilon
7	Armour Industrial Chemical Co. P.O. Box 1805 Chicago, Illinois USA	Arquad
8	Arron Chemicals Ltd.	Arfloc
9	Atlas Chemical Industries, Inc. Wilmington, Delaware 19899 USA	Sorbo
10	Berdell Industries 28-01 Thomson Ave. Long Island City, N.Y. 11101 USA	Berdell-Flocculant
11	F.W. Berk Ltd. Stratford London E 15, 3 NX England	Aripol
12	Betz Laboratories, Inc. 4636 Somerton Rd. Trevose, Pa. 19047 USA	Betz Polymer Poly-Floc 4D
13	Bond Chemicals Inc. 1500 Brookpark Rd. Cleveland, Ohio 44109 USA	Bondfloc
14	BTI Chemicals Ltd. 3 Chellow Street Manchester Rd., Bradford, Yorkshire England	BTI (Polymer)
15	Badische Anilin & Soda Fabrik AG B.A.S.F. Ludwigshafen am Rhein West Germany	Polimin Sedipur Polymin Luresin Polyflok Sursolan
16	Brennan Chemical Co. 704 North First St. St. Louis, Mo. 63102 USA	Brenco

Number	Manufacturer and/or Distributor (D)	Product
17	Brenco Corp. 704 N. First St. St. Louis, Mo. 63102 USA	Brenco
18	Buckman Lab., Inc. 1256 N. McLean Blvd. Memphis, Tenn. 38108 USA	Budond
19	Bulgaria	Uronoflok
20	The Burtonite Co. 49 Franklin Ave. P.O. Box 7 Nutley, N.J. 07110 USA	Burtonite
21	Calgon Corp. Water Chemicals Department P.O. Box 1346 Pittsburgh, Pa. 15222 USA	Calgon Coagulant Aid WT
21a	Calgon Corporation Polymer Department Water Management Division P.O. Box 1346 Pittsburgh, Pa. 15222 USA	Cat-Floc
21b	Chemviron SA (D) 13-15 Avenue des Arts Brussels, Belgium	Cat-Floc Polymer 225 Polymer 226 Polymer 233 Polymer 243 Polymer 253
21c	Chemviron Ltd. (D) 12 High Street Hampton Wick Kingston on Thames England	Cat-Floc
22	Cesalpinia SpA 24100 Bergamo Italy	Dealco
23	Commercial Chemical Products, Inc. 11 Patterson Ave. Midland Park, N.J. 07432 USA	Speedifloc
24	Crown Zellerbach Corp. 1 Bush St. San Francisco, Ca. 94119 USA	Orzan
25	Cyanamid G.B. Ltd. (D) Bush House Aldwych London W.C. 2 England	Aerofloc Cyanamer Magnifloc Superfloc

Number	Manufacturer and/or Distributor (D)	Product
25a	Cyanamid International Corp. (D) 8134 Adliswil Zürichstrasse Switzerland	Aerofloc Cyanamer Magnifloc Superfloc
26	Czechoslovakia	Flokal CMC Akrynax Polyakrylamid Starch Flokal Methamyl Akryflok
27	Dearborn Chemical Co. Div. W.R. Grace & Co. Merchandise Mart Plaza Chicago, Ill. 60654 USA	Aquafloc
28	Dow Chemical Company Barstow Bldg. 2020 Dow Center Midland, Michigan 48640 USA	Dow Dowell Purifloc Separan PEI SA-1704.2 XD-7817
28a	Dow Chemical GmbH (D) Concordiaplatz 3 Vienna 1 Austria	Purifloc
28b	Dow Chemical Co. (U.K.) Ltd. (D) 105 Wigmore Street London W. 1 England	Purifloc
29	EKOF-Erz Kohle Flotation West Germany	Ekofan
30	Electro Chemical Corp. P.O. Box 234 Lennox Hill Station New York, N.Y. 10021 USA	Ecco Suspension 　Catalyzer
31	Drew Chemical Corporation 701 Jefferson Rd. Parsippany, N.J. 07054 USA	Alchem Coagu-Aid Drewfloc DF-454 Amerfloc
32	DuBois Chemicals Div. of W.R. Grace & Co. DuBois Tower Cincinnati, Ohio 45202 USA	Flocculite

Number	Manufacturer and/or Distributor (D)	Product
33	E.I. Du Pont de Nemours & Co. Wilmington, Delaware 19898 USA	Carboxymethylcellulose
34	Environmental Pollution Investigation and Control Inc. 9221 Bond St. Overland Park, Kan. 66214 USA	Dyna Floc
35	Fabcon International 1275 Columbus Ave. San Francisco, Ca. 94133 USA	Fabcon Zuclar EZ-494
36	Farbwerke Hoechst AG (vorm. Meister Lucius & Brüning) 6230 Frankfurt am Main 80 West Germany	DT
37	VEB Fettchemie Karl Marx Stadt East Germany	Stipix
38	Henry W. Fink & Co. 6900 Silverton Ave. Cincinnati, Ohio 45235 USA	Kleer-Floc
39	Fisons Ltd. Agrochemical Division Harston Cambridge CB 2 - SH 4 England	Fi-clor
40	Float-Ore Ltd. Apex Works, Willowbanks Uxbridge, Middlesex England	Flocbel Flocgel
41	Forestal Industries U.K. Ltd. Ditton, Widness Lancashire England	Floccotan
42	Foseco Drayton Manor, Tamworth Staffordshire England	Decapol
43	Fospur Ltd. Drayton Manor, Tamworth Staffordshire England	Decapol

Number	Manufacturer and/or Distributor (D)	Product
43a	Fospur International Ltd. 36 Queen Anne's Gate London S.W. 1 H 9AR England	Decabloc
43b	Fospur Ltd. Alfreton Industrial Estate Nottingham Rd. Summercotes, Derbyshire England	Decapol
44	France	Actifloc Afcolac S100 Azyme Blanose Cellin Flocal Flocules CMC HX-19
45	Gamlen Chemical Co. Sybron Corporation 333 So. Victory Ave. San Francisco, Ca. 94080 USA	Gamlose W Gamlen Wisprofloc 20 Gamafloc NI-702
45a	Gamlen Chemie GmbH	Gamlose W
46	Garreth-Callaham 111 Rollins Road Millbrae, Ca. 94030 USA	Coagulant Aid—No Formula
47	GAF Corp. 140 West 51st St. New York, N.Y. 10020 USA	PVM/MA
48	General Mills Chemicals 4620 W. 77th Street Minneapolis, Minn. 55435 USA	Quartec Supercol Guar Gum
49	A.F. Goodman & Sons 21-07 41st Ave. Long Island City, N.Y. 11101 USA	Starches
50	B.F. Goodrich Chemical Co. 6100 Oak Tree Blvd. Cleveland, Ohio 44131 USA	Goodrite K
51	Hagan Chemicals & Controls, Inc. Pittsburgh, Pa. USA (see Calgon)	Hagan Coagulant Aid

Number	Manufacturer and/or Distributor (D)	Product
52	Hamada Servi Ind. Co., Ltd. Japan	Floc-Ace
53	Hercules Powder Co., Inc. 910 Market St. Wilmington, Del. 19899 USA	Ceron Cellulose-gum CMC Dresinate Natrosol Reten Hercofloc Carboxymethylcellulose
53a	N.V. Hercules Powder Company (D) P.O. Box 6189 The Hague The Netherlands	CMC Reten Hercofloc Delfloc
53b	Hercules Powder Co. (D) 1 Gt. Cumberland Place London England	Reten
54	Frank Herzl Corp. (D) 150 E. 58th St. New York, N.Y. 10017 USA	Perfectamyl
55	Messrs Avebe GA Veendam The Netherlands	Perfectamyl
56	Houseman & Thompson Ltd. The Priory Burnham, Bucks England	Zimmite Vaptreat
57	Chemische Fabrik Stockhausen & Cie 4150 Krefeld West Germany	Praestol
58	Chemische Fabrik Uetikon 8707 Uetikon am See Switzerland	Uetikon-flocculants
59	Chemviron SA P.O. Box 17 Ixelles 1 B 1050 Brussels Belgium	Polymer 233 Polymer 243 Polymer 253
60	ICI America, Inc. Wilmington, Del. 19899 USA	Atlasep
60a	ICI Ltd. P.O. Box 19, Templar House 81/87 High Holborn London W.C. 1 England	Cirragol Z

Number	Manufacturer and/or Distributor (D)	Product
61	Illinois Water Treatment Co. 840 Cedar St. Rockford, Ill. 61105 USA	Ollco IFA
62	Imperial Chemical Industries Ltd. P.O. Box 7, Winnington Northwich, Cheshire England	Alfloc DC Sedomax
63	Industrial Chemical Services 40 Heol Iscoed Llanishen Cardiff Wales	I.C.S.
64	Ionac Chemical Corporation Birmingham, N.J. 08011 USA	Ionac Coagulant Aid Ionac Wisprofloc
65	Japan	Accofloc Arafloc Sanpoly
66	J. & M. Polymers (Chemicals) Ltd. 5 Wellfield, Manor Farm Hazelmere, Bucks. England	JM 67 Jayfloc
67	Kelco Co. 75 Terminal Ave. Clark, N.J. USA	Kelcosol Kelgin Keltex
68	Kelco Co. 8225 Aero Drive San Diego, Ca. 92123 USA	Kelcosol Kelgin Keltex
69	Key Chemicals, Inc. 4346 Tacony St. Philadelphia, Pa. 19124 USA	Key-Floc
70	Kurito Industrial Comp. Ltd. Japan	Kurifloc
71	Lenning Chemicals Ltd. (D) 26-28 Bedford Row London S.W. 1 England	Primafloc
72	Hungary	Sedosan
73	Mayvil Chemicals Ltd. P.O. Box 23, Broad Heath Altrincham, Cheshire England	Katafloc

Number	Manufacturer and/or Distributor (D)	Product
74	L.C. Marquart GmbH Chemische Fabrik Siegburger Strasse 7 5300 Bonn-Beuel West Germany	Mafloc
75	Metalene Chemical Co. Bedford, Ohio 44014 USA	Metalene Coagulant
76	Meyhall Chemicals (U.K.) Ltd. Grosvenor House Market Parade Gloucester England	Algin Alginat Jaguar Polymer
77	Minoc (D) 17, Rue de Miromesnie Paris 8^0 France	Primafloc
78	Monsanto Chemical Co. Ind. Resins Department Plastics Division Springfield, Mass. USA	DX-908 Krilium Lustrex Reagent RD Lytron
79	Monsanto Europe SA (D) 1 Place Madon Brussels 3 Belgium	Krilium Lytron
80	Nalco Italiano SpA (D) Viale dell'Esperanto 71 00144 Rome Italy	Nalcolyte Nalco
81	Nalco Chemical Co. 180 N. Michigan Chicago, Ill. 60601 USA	Nalco Nalcolyte
82	Nalfloc Limited P.O. Box 11 Northwich Cheshire England	Nalfloc R 177
83	Narvon Mining & Chem. Co. Box 1598-T Lancaster, Pa. 17604 USA	Sink-Floc Zeta-Floc

Number	Manufacturer and/or Distributor (D)	Product
84	National Starch & Chemical Corp. 1700 West Front St. Plainfield, N.J. 07063 or 750 Third Ave. New York, N.Y. 10017 USA	Floc-Aid Starches 613-45 Natron Resyn 3285
85	Nitrokemia Hungary	Floconit
86	Mogul Corp. 20600 Chagrin Blvd. Cleveland, Ohio 44122 USA	Mogul Mogul Claracel Co. Ecco Suspension- 　Catalyzer
87	Oakite Products, Inc. 50 Valley Road Berkeley Hts., N.J. 07922 USA	Clarifier Enprox
88	O'Brien Industries, Inc. 513 W. Mt. Pleasant Ave. Livingston, N.J. 07039 USA	O'B-Floc
89	Oxford Chemical Div. Consolidated Foods Corp. P.O. Box 80202 Atlanta, Ga. 30341 USA	Oxford Hydrofloc
90	Petrolite Tretolite Div. 369 Marshall St. St. Louis, Mo. 63119 USA	Tretolite Tolfloc
91	Pfaudler International GmbH (D) St. Albangraben 3 Basel 10 Switzerland	Ionac
92	Poland	Gigamal Gigamid Gigtar Gipan Hypan P-25
93	Reichhold Chemicals 523 N. Broadway White Plains, N.Y. 10602 USA	Aquarid

Number	Manufacturer and/or Distributor (D)	Product
94	Rohm & Haas Company Independence Mall West Philadelphia, Pa. 19105 USA	CW Dansil DF-381 Primafloc Rohafloc
95	Sanyo Chemical Japan	Sanfloc
96	N.V. Scholten's Chemische Fabrieken N.V. Foxhol 8321 The Netherlands	Wisprofloc Flocgel 1297
97	Dr. Theodor Schuchardt & Co. GmbH Postfach 80 1549 8, Munich 80 Germany	Galactasol
98	Soviet Union	AMF GF IM-100 METAS Type K KODT PANG Polyakrylamid SP VA
99	Standard Brands Chemical Industries, Inc. 625 Madison Ave. New York, N.Y. 10022 USA	Tychem
100	A.E. Staley Mfg. Co. 2200 E. Eldorado Decatur, Ill. 62521 USA	Hamaco
101	Stein Hall & Co., Inc. 605 Third Ave. New York, N.Y. 10016 USA	Hallmark Jaguar Polyhall Polymer F-3 Reagent MRL, F, G, H Starches
102	Swift Chemical Company 115 W. Jackson Blvd. Chicago, Ill. 60604 USA	Swift's Polymer
103	Synthomer Chemie GmbH Postfach 3724 6000 Frankfurt am Main 1 West Germany	Synthofloc

Number	Manufacture and/or Distributor (D)	Product
104	Sweden (Alfa-Loval)	Cellufixa Separator PX
105	Taiwan	PMA-848
106	TAR Residuals Ltd. & Simon Engineering Co. (D) Cheadle Heath, Stockport Cheshire SK 3 ORY Plantation House, Mincing Lane London E.C. 3 England	Gelfloc Polyflok Powdaflok Tragaflok Wisprofloc Filtaflok TR-Superfloc
106a	The Midland Tar Distillers Ltd. Four Ashes Wolverhampton England	RD 2180 RD 2181
107	The Permutit Co. 49 East Midland Ave. Paramus, N.J. 07652 USA	Permutit
108	Union Carbide Corp. 270 Park Ave. New York, N.Y. 10017 USA	Polymer Polyox U.C. 149 Ucar Resin
109	James Varley & Sons, Inc. 1200 Switzer Ave. St. Louis, Mo. 63147 USA	Varco-Floc
110	VEB Chemische Werke Buna Schkopau über Merseburg East Germany	Verdickung AN
111	Wilkinson and Simpson Ltd. Low Friar Street Newcastle upon Tyne England	Vaporin
112	Yorkshire Dyeware and Chemical Co., Ltd. 27 Kirkstaff Rd. Leeds, Yorks. LS 31h England	Gelfloc Polyflok Powdaflok Tragaflok Wisprofloc Filtaflok
113	Zimmite Corp. 810 Sharon Drive Westlake, Ohio 44145 USA	Zimmite ZT-600 to 605 ZT-650
114	Carus Chemical Co., Inc. 1500 Eighth St. La Salle, Ill. 61301 USA	Caraflok

WATER-SOLUBLE RESINS
AND POLYMERS 1976
Technology and Applications

by Yale L. Meltzer

Chemical Technology Review No. 57

An intense competitive struggle for markets has developed among water-soluble synthetic polymers, the semisynthetic cellulose ethers and starch derivatives, and natural gums.

Modified starches hold an advantage over synthetics and natural gums, because they are less expensive. Starch products, however, can cause stream pollution in various industrial applications, while the true synthetics, with their lower BOD, produce considerably less polluting effluents.

Strongly favoring the synthetics, moreover, is the great versatility which they have displayed in their applications. In addition, the industrialized countries have developed elaborate marketing and product information systems for these substances.

Production of the natural gums is concentrated in the developing countries of Africa and Asia. Government officials there have expressed grave concern over the loss of markets to the synthetics.

Starch is obtained in quantity in many parts of the world. Corn starch holds the biggest share and the USA is its largest producer.

Over 200 patent-based manufacturing and end-use processes are described in this book. What makes it most valuable are the market survey data that precede the technological information. A much shortened, partial table of contents follows here. Chapter headings and some important subtitles are given. Numbers in parentheses indicate the number of processes per topic.

INTRODUCTION
Broad View of the Market
Main Uses and Applications

1. MARKET SURVEY
Wherever available the market survey portion contains a breakdown by
Applications
Producers and Capacities
Production, Sales, Prices
Market Growth Forecasts for each
  class of substances

2. ACRYLAMIDE POLYMERS (20)
Substituted Polyacrylamides
N-3-Aminoalkyl Polyacrylamides
In Visoelastic Fluids

3. ACRYLICS & METHACRYLICS (22)
Improved Reppe Process
Siloxane-Acrylate Copolymers

4. CELLULOSE ETHERS (13)
Partial Cellulose Ethers
Graft Copolymers

5. Na-CARBOXYMETHYLCELLULOSE (14)
High Viscosity CMC
Cement Compositions
Improved Process for Drilling Muds

6. HYDROXYETHYLCELLULOSE (8)
Filterable Solutions
Suspension Polymerization Process

7. HYDROXYPROPYLCELLULOSE (6)
For Biodegradable Containers

8. METHYL CELLULOSE (3)

9. ETHYL CELLULOSE (3)

10. ETHYLENE OXIDE POLYMERS (12)
Water-Dispersible Hot Melts
Fiber Lubricating Composition

11. POLYAMIDE RESINS (4)
Soluble Amide-Imide Polymers

12. POLYETHYLENIMINES (22)
Water-Soluble Copolymers
Addition Polymers
In Paper Sizes

13. POLYVINYL ALCOHOL (30)
Production Processes
Application Processes

14. POLYVINYL PYRROLIDONE (21)
Functional Polymers
Gelling Organic Polar Liquids
In Cosmetics and Pharmaceuticals
Producing Dye-Receptive Filaments

15. MODIFIED STARCHES (31)
Hydrolysates
Thixotropic Starch
Desolventizing Starch
Pregelatinized Starch Products
Starch Conversion Products
Phosphoric Acid Derivatives
Sulfite-Carbonyl Starch Complex
Inhibited Granular Starch Bases
Starch Phosphonate
Starch in Beer Production

16. NATURAL GUMS (4)
Modified Natural Gums
Processing of Macrocystis Kelp
Quinidine Alginate
Suspension Polymerization
  in the Presence of Xanthan Gum

17. STANNIC OXIDE POLYMERS (1)

ISBN 0-8155-0601-5

371 pages

INDUSTRIAL WATER PURIFICATION 1974

by Louis F. Martin

Pollution Technology Review No. 16

The Federal Water Pollution Control Act (now made into law) will have far-reaching effects on all industry. The pretreatment standards, about to be published as a result of this act, will exert increasing constraints on industrial process design and expansion plans in the years ahead.

The *Guidelines for the Pretreatment of Discharges of Publicly-Owned Works* prohibit the introduction of industrial pollutants which would pass through a municipal facility "inadequately treated," therefore, all existing industrial facilities and new plant designs must include specific downstream water purification processes, before the effluent is sewered or discharged into public systems.

This book describes over 160 recently devised processes for the treatment of contaminated industrial waters.

A partial and condensed table of contents follows here. Numbers in parentheses indicate the number of processes per topic, chapter headings are given, followed by examples of important subtitles.

While conventional separation processes for solid-liquid and water-oil effluents are applicable to many industrial streams, the process literature clearly highlights the areas of greatest immediate concern as shown by this table of contents:

1. SOLID-LIQUID SEPARATION PROCESSES (29)

Flocculation
Methylamine-Epichlorohydrins
Polyelectrolytes
Coagulation and Clarification
Continuous Filtration
Other Separation Techniques
Ion Exchange
Electrostatic and
    Electromagnetic Methods
Suspended Solids Techniques

2. OIL-WATER SEPARATIONS (18)

Special Equipment Needs
Flotation Cells
Gravity Separation
Vortex Separator
Multiphase Liquid/Liquid Separations
Use of Absorbents
Polypropylene Oil Mops
Permeable Foams for Coalescence

3. REMOVAL OF METALS (30)

Separation of Heavy Metals
Mercapto-s-Triazines
N-Acylamino Acids
Ion-Selective Membranes
Electrolytic Methods
Removal of Metal Ions
Specific Mercury Removal Methods
Other Specific Processes
Nuclear Fuel Wastes
Metal-Containing Lubricants
Organic Lead Removal
Vanadium and Cobalt Removal

4. METAL FINISHING EFFLUENTS (18)

Cyanide Removal
Conversion to Non-Toxic
    Biodegradable Compounds
Chromium Removal
Control of Aluminum
Other Removal Processes
Use of Chelates

5. PULP AND PAPER PROBLEMS (18)

Closed Cycle Process for
    Treating Waste Liquors
Hydropyrolysis and Use of
    Activated Carbon
Wet Air Oxidation
Flocculation
Oxyaminated Polyacrylamides
Sulfur Recovery
Ultrafiltration

6. COAL, ORE AND TAR SAND PROCESSING (14)

Reverse Osmosis + Neutralization
Use of Waste Steam
Salt-Free Condensates
Use of Nucleoproteins
Cyclone Floats

7. REFINING and CHEMICAL PROCESSES (34)

Refinery Operations
Fertilizers
Paint
Phenol
Photography
Meat Processing
Other Specific Processes

ISBN 0-8155-0554-X

300 pages

CORROSION RESISTANT MATERIALS HANDBOOK 1976

Third Edition

by Ibert Mellan

This famous book, first published in 1966 and now already in its third, greatly enlarged and completely revised edition, will help you cut losses due to corrosion by enabling you to choose the proper *commercially available* corrosion resistant materials for your particular purpose.

The great value of this outstanding reference work lies in the extensive cross-indexing of thousands of substances. The 151 tables are arranged by types of **corrosion resistant materials.** The **Corrosive Materials Index** is organized by *corrosive chemicals* and other *corrosive substances:* it refers you to specific recommendations in the tables. New in this edition are a separate **Trade Names Index** and a listing of **Company Names and Addresses.**

For the first time there appears also a group of 13 tables comparing the respective anticorrosive merits of commercial engineering and construction materials essential to industry.

The tables in this book represent selections from manufacturers' literature made by the author at no cost to, nor influence from, the makers or distributors of these materials.

Contents:

1. SYNTHETIC RESINS AND POLYMERS (87 Tables)

2. RUBBERS AND ELASTOMERS (8 Tables)

3. CEMENTS AND MORTARS (13 Tables)

4. METALS AND ALLOYS (27 Tables)

5. GLASS AND CERAMICS (2 Tables)

6. WOOD (1 Table)

7. COMPARATIVE CORROSION RESISTANCES OF MATERIALS OF CONSTRUCTION (13 Tables)

8. CORROSIVE MATERIALS INDEX

9. TRADE NAMES INDEX

10. COMPANY NAMES AND ADDRESSES

ISBN 0-8155-0628-7

665 pages

HOW TO DISPOSE OF TOXIC SUBSTANCES AND INDUSTRIAL WASTES 1976

by Philip W. Powers

Environmental Technology Handbook No. 4

This book discusses all recognized ultimate disposal methods in detail and contains a long list of specific recommendations for specific substances plus alternative disposal or recovery methods.

Ultimate waste disposal implies the final disposition of non-degradable, persistent, harmful, and cumulative wastes that may be solid, liquid, or gaseous. Workable solutions to ultimate disposal problems as described in this book include conversion to harmless end products, subsurface storage in ponds or landfills, and disposal in the ocean. This condensed information will enable anyone to establish a sound background for action towards disposal of toxic and hazardous materials with safety.

The five general categories of hazardous wastes are:

1. Toxic Chemical
2. Radioactive
3. Flammable
4. Explosive
5. Toxic Biological

Many toxic or hazardous wastes contain valuable materials. Whenever this is the case, recovery and reuse is one of the most desirable methods of hazardous waste avoidance.

The book is based on government-sponsored reports (often elaborated by highly qualified personnel from industrial companies), U.S. patents, and on pertinent articles in authoritative journals. A partial and condensed table of contents follows here. Chapter headings are given in full whenever possible.

INTRODUCTION
  Economics of Ultimate Disposal

1. PRETREATMENTS
  Cementation
  Sand Drying Beds
  Submerged Combustion

2. MAJOR DISPOSAL METHODS
  Biological Degradation
  Chemical Degradation (Chlorination)
  Composting
  Long Term Storage
  Oxidation and Pyrolysis

3. DEEP WELL DISPOSAL
  Effluent Compatibility
  Legalities and Economics

4. INCINERATION
  Catalytic Incinerators
  Fluidized Bed Incinerators
  Seagoing Incinerators

5. LAND APPLICATION

6. LAND BURIAL

7. LANDFILLS
  Regulatory Considerations

8. OCEAN DISPOSAL
  Basic Procedures

9. AUTOMOTIVE INDUSTRY WASTES

10. ELECTRIC INDUSTRY WASTES

11. FOOD INDUSTRY WASTES
  Animal Wastes
    Pyrolysis of Fish Wastes
  Exhaust Fume Treatments

12. GRAPHIC ARTS WASTES
  Printing Process Effluents

13. HOSPITAL WASTES

14. INORGANIC CHEMICAL WASTES
  Ammonia Plant Effluents
  Asbestos Problems
  Chrome Pigments & Lead
  Cyanides & Hydrazines

15. IRON & STEEL WASTES
  Coke Oven & Furnace Effluents
  Foundry Wastes

16. METAL FINISHING WASTES

17. NUCLEAR WASTES
  Spent Fuel Disposal
  Waste Isotope & Tritium Disposal

18. ORDNANCE INDUSTRY WASTES
  Chemical Warfare Agents
  Explosives Disposal

19. ORGANIC CHEMICAL WASTES
  Chlorinated Hydrocarbons
  Mercaptans
  Nitro Compounds

20. PAINT INDUSTRY WASTES

21. PESTICIDE WASTES

22. PETROLEUM WASTES

23. DRUG INDUSTRY WASTES

24. PULP & PAPER WASTES

25. RUBBER & PLASTIC WASTES

26. TEXTILE & WOOD WASTES

APPENDIX: Names & Addresses
  of Disposal Contractors

ISBN 0-8155-0615-5

500 pages

RESOURCE RECOVERY
AND
RECYCLING HANDBOOK
OF INDUSTRIAL WASTES
1975

by Marshall Sittig

Environmental Technology Handbook No. 3

This handbook intends to show the conversion of industrial wastes into resources. Increasing populations and ever expanding industrialization have increased waste output to a point where natural reclamation can no longer keep up. The synthesis pathways of nature are overloaded.

Wastes in the ecological sense are manufactured goods no longer useful to man, as are by-products and residua (dross, slag, offal) as well as liquid and gaseous effluents, including raw sewage.

This volume concentrates on the process technology available for resource recovery and recycling of industrial wastes. Financial and economic considerations and allied decisions must be left to the potential users of such processes, since their practicality varies widely with local conditions and the shortage or over-abundance of raw materials.

As in our other handbooks on pollution control, vital data are condensed from U.S. Government-sponsored projects and recent U.S. patents. The following gives important highlights from the large table of contents. Wherever possible, entries have been arranged in an alphabetical and encyclopedic manner. Numbers in parentheses indicate numbers of recovery processes per topic.

1. COMPLEX WASTE PRODUCTS (20)
Scrap Automobiles
Sewage
Useful Products from Sewage
Fertilizers from Activated Sludge

2. FERROUS METALS (16)
From Riverbed Slimes
From Cans and Drums
From Alloy Scrap
From Slag
From Phosphatizing Sludge

3. FOOD & BEVERAGE WASTES (18)
For Animal Feeds
Proteins from Whey
Proteins from Animal Wastes
Fertilizers from Citrus Pulp
Tallow from Offal

4. GLASS AND CERAMICS (6)
Separation from Municipal Waste
Use in Brick Manufacture
Use in Cement
Recovery from Grinding
Regeneration of Foundry Sand

5. HEAT RECOVERY (6)
From Waste Gases
From Nuclear Reactors
From Power Plant Coolants
From Municipal Incinerators
From Industrial Incinerators

6. INORGANIC MATERIALS (57)
Ammonia Recovery
Fluorine and Fluoride Recovery
Phosphate Recovery
Sulfur Compounds Recovery
Titanium from Paper Mill Sludge

7. NATURAL PRODUCTS (12)
Charcoal from Bagasse
Cattle Feed from Banana Peels
Fertilizer from Castor Bean Residues
Leather from Leather Scrap
Furfural from Oat Hulls

8. NONFERROUS METALS (102)
Aluminum Recovery
Cadmium from NiCad Batteries
Cobalt from Nickel Ore Tailings
Germanium from Zinc Smelting
Mercury from Aqueous Wastes

9. ORGANIC MATERIALS (34)
Recovery of Chlorinated Solvents
Ethylene Glycol from Polyester Wastes
Hydrocarbons from Flare Systems
Recovery of Used Motor Oil
Starch from Potato Processing
Vanillin from Sulfite Liquors

10. PAPER & WOOD MILL WASTES (26)
Soil Conditioner from Bark
Recovery of Sugars
Recovery of Single Cell Protein
Paper Recycling Situation
Fuel Gas from Kraft Liquor

11. PLASTICS AND FIBERS (20)
Polyesters from Spinning Wastes
Polyesters from Used Photo Films
Pyrolysis to Recover Monomers
Plasticizer Recovery by Extraction
Polyvinyl Chloride by Extraction

12. RUBBER RECLAIMING (12)
Depolymerization Processes
Mineral Values from Scrap Rubber
Resins from Scrap Rubber
Recovering Rubber by Hydroconversion

13. PROJECTIONS & FUTURE TRENDS

ISBN 0-8155-0552-2

427 pages

WASTEWATER CLEANUP
EQUIPMENT 1973

Second Edition

Water pollution is becoming more of a problem with every passing year. Plants engaged in all types of manufacture are being more and more carefully watched by federal, state, and municipal governments to prevent them from pouring their untreated effluents into the nation's waterways, as they used to do. The sewage treatment plants of many municipalities are becoming too small for the burgeoning population, and many communities once served by individual septic tanks are having to build sewers and treatment plants.

Water pollution will be solved primarily by application of techniques, processes, and devices already known or in existence today, supplemented by modifications of these known methods based on advanced technology. This book gives you basic technical information and specifications pertaining to commercial equipment currently available from equipment manufacturers. Altogether the products of 94 companies are represented.

This second edition of "Wastewater Cleanup Equipment" supplies technical data, diagrams, pictures, specifications and other information on commercial equipment useful in water pollution control and sewage treatment. The data appearing in this book were selected by the publisher from each manufacturer's literature at no cost to, nor influence from, the manufacturers of the equipment.

It is expected that vast sums will be spent in the United States during the remaining portion of this decade for control and abatement of water pollution. Much of the expenditure will be for the type of equipment described in this book.

Today's environmental control is taken to mean a specialized technology employing specialized equipment designed to process the discarded and excreted wastes of human metabolism and human activity of any sort.

Next to air, water is the most abundant and utilized commodity necessary for the maintenance of human life. The average consumption of water per person in residential communities in the United States is between 40 and 100 gallons in one day. In highly industrialized communities the average consumption pro head can be as high as 250 gallons per day.

The reuse of wastewater after cleanup is not only becoming a cogent necessity, but it is also becoming more attractive economically. The degree of purity required for industrial water use is in many cases greater or vastly different from that acceptable for potable water.

Special equipment for cleanup of wastewater is therefore an absolute must, and this book is offered with the intention of providing real help in the selection of the proper equipment.

The descriptions and illustrations given by the original equipment manufacturer include one or more of the following:

1. **Diagrams of commercial equipment with descriptions of components.**

2. **A technical description of the apparatus and the processes involved in its use.**

3. **Specifications of the apparatus, including dimensions, capacities, etc.**

4. **Examples of practical applications.**

5. **Graphs relating to the various parameters involved.**

Arrangement is alphabetically by manufacturer. A detailed subject index by type of equipment is included, as well as a company name cross reference index.

ISBN 0-8155-0487-X

372 pages

POLLUTION DETECTION

AND MONITORING HANDBOOK 1974

by Marshall Sittig

Environmental Technology Handbook No. 1

This handbook contains methods for the detection and monitoring of pollutants in industrial effluents, notably air and water.

Such methods are prerequisites for any kind of management of environmental quality. Of equal importance are the data handling systems that must be used to collect and interpret the measurements, regardless of whether manual or automated identifying and monitoring techniques are chosen.

Generally the most important single factor in determining whether manual or automated methods of analysis and level monitoring should be used, is the required frequency interval. In the manual approach costs vary almost directly with the measurement frequency. At monitoring intervals oftener than once a day, or when a continuous watch must be maintained, installation of automatic sensing is more economical, provided reliable automatic sensors are available.

For meaningful control and prevention of pollution the accepted standards and parameters for air and water quality must be known, whether they have been legislated or not. Acceptable levels for each type of impurity from methylmercury to dissolved chlorides or carbon particles must be known by the industrialist or public health official, and this book, with its 1633 references makes a massive attempt to communicate such levels.

As a guideline for the industrial user, data on the toxicity of the various pollutants are included. Such data on the deleterious effects were taken from various government publications. This book is not a treatise of pharmacology or industrial toxicology, but a guide to the accurate evaluation of pollutant levels in a given effluent or ambient substrate.

Whenever applicable, the name of the pollutant is followed by chemical and other information. After this come directions for sampling in air or water or both, as the case may be. Thereafter are given qualitative and quantitative analytical methods suitable for identification and measurement of the degree or level of pollution. This comprehensive information is followed by measurement techniques suitable for repeated or continuous monitoring. Preference is given to methods giving reproducible results in environmental quality control, as recommended by the EPA and other U.S. government agencies. This is followed in each case by many up-to-date references to government publications, patents, and journal articles.

Arrangement is encyclopedic. The book is thus a worthy companion to the author's POLLUTANT REMOVAL HANDBOOK available from the same publisher.

Pollution control is a serious business, and in the design of a monitoring program the specific objectives to be served must be clearly in mind from the very beginning. Otherwise the effort most probably will not result in an efficient and meaningful program. The length of the initial survey, selection of monitoring points, parameter coverage, and sampling frequency all depend on the specific objectives and the timeframe in which the objectives must be achieved. These variables also have a very significant impact on monitoring costs.

Monitoring for the purpose of long-term trend identification, evaluation of standards compliance, and total management of environmental quality must, of course, be a continuing program without end.

It is hoped sincerely that this book, together with its companion volume POLLUTANT REMOVAL HANDBOOK, will do much to establish and sustain such antipollution programs.

A partial and condensed table of contents follows here in which subheadings are indicated below for only the first few entries. The book contains a total of 88 subject entries arranged in an alphabetical and encyclopedic fashion. The subject name refers to the polluting substance, and the text underneath each entry tells how to measure and monitor pollution by said substance:

INTRODUCTION

DEFINITION OF PROBLEMS
  Terms Used

POLLUTANT EFFECTS
  On Health
  On the Environment

AMBIENT QUALITY STANDARDS

EMISSION STANDARDS
  For Stationary Sources
  For Mobile Sources

MONITORING

ANALYTICAL SYSTEMS
  Manual
  Automated (General)
  Automated (Air)
  Automated (Water)
  Remote
  Public

This book should prove to be very useful as an overall reference volume. The organization is clear, and the text is especially well arranged. The tables and figures are extensive and will be of value to students, engineers, and pollution control authorities.

ISBN 0-8155-0529-9

401 pages

ENVIRONMENTAL SOURCES
AND EMISSIONS HANDBOOK 1975

by Marshall Sittig

Environmental Technology Handbook No. 2

Environmental pollution with its far-reaching effects is only now beginning to be understood. In this practical handbook the various modifications a given pollutant can assume, are traced back to their origins, and the intermedia transfers are discussed fully. Intermedia transfers include direct transfer (removal of a pollutant from one medium and its disposal in another) or indirect (pollution created in another medium and usually in another form by a basic change in a process or industry).

Also, an example of insidious conversion is mercury. Metallic mercury is relatively innocuous, but when organic molecules or dead organisms are present in rivers and lakes, mercury reacts with these molecules to form toxic methylmercury compounds which are excreted very slowly by fish and man.

The implications are formidable: Chemical, biochemical, technological, statistical, mineralogical, zoological, pharmacological, medical, legal, legislative and public health involvements are all too obvious.

This volume therefore surveys the origins of both air and water pollution. Significant emphasis is placed on altered pollution which can result when an air pollutant is intentionally or accidentally transferred to an aqueous stream or vice versa.

Sometimes pollutants react with each other or initiate deleterious chain reactions. Yet this same reactivity may be the key to efficient removal: by adding certain chemicals or flocculants or microorganisms, toxic substances may be carried off physically or converted not only to nontoxic substances, but also into useful products.

The why and where is in this book, commensurate with present-day technology, without becoming too sophisticated or losing sight of the all important economic considerations.

Operators or potential operators of processes which produce pollutants will find this volume quite useful. Besides discussing all sorts of pollution sources, it should help to define industry-wide emission practices and magnitudes.

A partial and condensed table of contents follows here. The book contains 200 subject entries and 382 tables with figures and graphs. Descriptions are mostly based on studies conducted by industrial and engineering firms or university research teams under the auspices of various government agencies. As is the case with other handbooks in this series the entries are arranged in an alphabetical and encyclopedic fashion:

**SOURCES OF SPECIFIC
  TYPES OF POLLUTANTS**
Acids & Alkalis
Aldehydes
Ammonia
Arsenic
Asbestos
Barium
Beryllium
Biological Materials
Boron
Cadmium
Carbon Monoxide
Chlorides
Chlorine
Chromium
Copper
Cyanides
Ethylenes
Fluorides
Heavy Metals
Hydrocarbons
Hydrogen Chloride
Hydrogen Cyanide
Hydrogen Sulfide
Iron
Lead
Magnesium
Manganese
Mercury
Molybdenum
Nickel
Nitrogen Compounds
Nitrogen Oxides
Odorous Compounds
Oily Wastes
Organics
Particulates
Pesticides
Phenols
Phosphorus (Elemental)
Phosphorus Compounds
Pollens
Polynuclear Aromatics
Radioactive Materials
Selenium
Silver
Solid Waste Incineration Products
Sulfur Oxides
Surfactants
Suspended Solids
Tin
Titanium Dioxide
Total Dissolved Solids
Vanadium
Zinc
EMISSIONS FROM SPECIFIC PROCESSES
Adipic Acid Manufacture
Aircraft Operation
Alfalfa Dehydrating
Aluminum Chloride Manufacture

Aluminum Production, Primary
Aluminum Industry, Secondary
Aluminum Sulfate Manufacture
Ammonia Manufacture
Ammonium Nitrate Manufacture
Ammonium Sulfate Production
Asbestos Products Industry
Asphalt Roofing Manufacture
Asphaltic Concrete Plants
   (Asphalt Batching)
Automobile Body Incineration
Automotive Vehicle Operation
Bauxite Refining
Beet Sugar Manufacture
Boat & Ship Operation
Brass & Bronze Ingot Prod.
   (Secondary Copper Industry)
Brick & Related Clay
   Products Manufacture
Builders Paper Manufacture
Calcium Carbide Manufacture
Calcium Chloride Manufacture
Carbon Black Manufacture
Castable Refractory Manufacture
Cement Manufacture
Ceramic Clay Manufacture
Charcoal Manufacture
Chlor-Alkali Manufacture
Clay & Fly Ash Sintering
Coal Cleaning
Coal Combustion, Anthracite
Coal Combustion, Bituminous
Coffee Roasting
Coke (Metallurgical) Manufacture
Combustion Sources
Concrete Batching
Conical Burners
Copper (Primary) Smelting
Cotton Ginning
Dairy Industry
Dry Cleaning
Electroplating
Explosives Manufacture
Feedlots
Fermentation Industry
Ferroalloy Production
Fertilizer Industry
Fiberglass Manufacture
Fish Processing
Frit Manufacture
Fruit & Vegetable Industry
Fuel Oil Combustion
Glass Manufacture
Gold & Silver Mining & Production
Grain & Feed Mills & Elevators
Gypsum Manufacture
Hydrochloric Acid Manufacture
Hydrofluoric Acid Manufacture
Hydrogen Peroxide Manufacture
Incineration, Municipal Refuse
Inorganic Chemical Industry
Iron Foundries
Iron & Steel Mills
Lead Smelting, Primary
Lead Smelting, Secondary
Leather Industry
Lime Manufacture
Liquefied Petroleum Gas Combustion
Magnesium Smelting, Secondary

Meat Industry
Mercury Mining & Production
Mineral Wool Manufacturing
Natural Gas Combustion
Nitric Acid Manufacture
Nitrogen Fertilizer Manufacture
Nonferrous Metals Industry
Open Burning
Orchard Heating
Organic Chemicals Manufacture
Paint & Varnish Manufacture
Paperboard Manufacture
Perlite Manufacture
Pesticide Manufacture
Petroleum Marketing & Transportation
Petroleum Refining
Petroleum Storage
Phosphate Chemicals Manufacture
Phosphate Fertilizer Manufacture
Phosphate Rock Processing
Phosphoric Acid Manufacture
Phosphorus Manufacture
Phosphorus Oxychloride Manufacture
Phosphorus Pentasulfide Manufacture
Phosphorus Pentoxide Manufacture
Phosphorus Trichloride Manufacture
Phthalic Anhydride Manufacture
Plastics Industry
Potassium Chloride Production
Potassium Dichromate Manufacture
Potassium Sulfate Manufacture
Poultry Processing
Power Plants
Printing Ink Manufacture
Pulp & Paper Industry
Railway Operation
Refractory Metal (Mo, W) Production
Rubber Industry, Synthetic
Sand & Gravel Processing
Seafood Processing Industry
Sewage Sludge Incineration
Soap & Detergent Manufacture
Sodium Bicarbonate Manufacture
Sodium Carbonate Manufacture
Sodium Chloride Manufacture
Sodium Dichromate Manufacture
Sodium Metal Manufacture
Sodium Silicate Manufacture
Sodium Sulfite Manufacture
Starch Manufacture
Stationary Engine Operation
Steel Foundries
Stone Quarrying & Processing
Sugar Cane Processing
Sulfur Production
Sulfuric Acid Manufacture
Surface Coating Application
Synthetic Fiber Manufacture
Terephthalic Acid Manufacture
Textile Industry
Timber Industry
Tire & Inner Tube Manufacture
Titanium Dioxide Manufacture
Urea Manufacture
Wood Veneer & Plywood Products Industry
Wood Waste Combustion
Zinc Processing, Secondary
Zinc Smelting, Primary

ISBN 0-8155-0568-X   521 pages

TP
159
F54
V67

Vostrcil, Josef.

Commercial organic
  flocculants

DATE			